Introduction to PHYSICAL HYDROLOGY

EDITED BY

Richard J Chorley

CONTRIBUTORS
R. G. Barry, R. P. Beckinsale, M. A. Carson, R. J. Chorley
R. W. Kates, M. J. Kirkby, M. G. Marcus, R. J. More,
J. C. Rodda, D. B. Simons, and J. P. Waltz

METHUEN & CO LTD

First published in 1969
First published as a University Paperback in 1971
Reprinted twice
Reprinted 1977

Printed in Great Britain by
Richard Clay (The Chaucer Press), Ltd
Bungay Suffolk

ISBN 0 416 68810 1

Distributed in the USA by
HARPER & ROW PUBLISHERS, INC.
BARNES & NOBLE IMPORT DIVISION

Contents

Preface to the Paperback Edition

This paperback originally formed part of a larger, composite volume entitled *Water, Earth, and Man* (Methuen and Co Ltd, London, 1969, 588 pp.), the purpose of which was to provide a synthesis of hydrology, geomorphology, and socio-economic geography. The present book is one of a series of three paperbacks, published simultaneously, which set out these themes separately under the respective titles:

Introduction to Physical Hydrology
Introduction to Fluvial Processes
Introduction to Geographical Hydrology

The link with the parent volume is maintained by the retention of the Introduction, which gives the rationale for associating the three themes. The aim of this paperback is primarily to make available in a cheap and handy form one of these systematic themes. In doing so, however, it is hoped that the book will provide a constant reminder of the advantages inherent in adopting a unified view of the earth and social sciences, and, in particular, that the study of water in the widest sense presents one of the most logical means of increasing our understanding of the interlocking physical and social environments.

Acknowledgements

The editor and contributors would like to thank the following editors, publishers, and individuals for permission to reproduce figures and tables:

Editors

American Journal of Science for fig. 2.II.8; *Bulletin of the American Association of Petroleum Geologists* for fig. 2.II.15; *Bulletin of the Geological Society of America* for figs. 2.II.4, 2.II.11, 2.II.17, and 5.I.6(*a*); *Bulletin of the International Association of Scientific Hydrology* for figs. 4.II.3, 4.II.5, and 5.I.3; *Department of Civil Engineering Technical Reports, Stanford University* for figs. 2.I.2 and 9.I.5; *Geographical Bulletin* for fig. 8.I.5(*a–c*); *Journal of Applied Meteorology* for fig. 3.I(i).3; *Journal of Geology* for fig. 2.II.18; *Journal of Glaciology* for fig. 8.I.3; *Petermann's Geographische Mitteilungen* for fig. 1.I.4; *Proceedings of the Institution of Civil Engineers* for fig. 9.I.3; *Transactions of the American Geophysical Union* for figs. 2.I.3 and 2.II.20.

Publishers

Edward Arnold Ltd., London for fig. 4.I.4 from *Irrigation and Climate* by H. Oliver; The Macmillan Co., New York, for fig. 11.I.1 from *Rainfall and Runoff* by E. E. Foster; McGraw-Hill Book Co., New York, for figs. 2.II.3, 2.II.6, and 2.II.14 from *Handbook of Applied Hydrology* by Ven Te Chow (Ed.); Methuen and Co. Ltd., London, for fig. 1.I.1 from *Models in Geography* by R. J. Chorley and P. Haggett (Eds.); Thos. Nelson and Sons, London, for figs. 3.I(i).4 and 11.I.5 from *The British Isles* by J. W. Watson and J. B. Sissons (Eds.); University of Chicago Press for figs. 1.I.5, 1.I.9, and 4.I.7 from *Physical Climatology* by W. D. Sellers; John Wiley and Sons, Inc., New York, for fig. 2.II.13 from *Geohydrology* by R. J. M. DeWiest; U.N.E.S.C.O. for fig. 8.I.3.

Individuals

The Controller, Her Majesty's Stationery Office (Crown Copyright Reserved) for fig. 1.I.3; The Director, Geographical Branch, Office of Naval Research, Washington, for fig. 2.II.16 from the *Technical Report* by A. Broscoe and fig. 2.II.21 from the *Technical Report* by M. A. Melton; The Director, Military Engineering Experiment Establishment, Christchurch, Hampshire, for fig. 2.II.1; The Director, Tennessee Valley Authority, Office of Tributary Development, Knoxville, Tennessee, for fig. 5.I.5 from the *Research Paper* by R. P. Betson;

The Director, U.S. Army Engineer Experiment Station, Vicksburg, Mississippi, for fig. 2.11.1; The Director, U.S. Geological Survey for figs. 2.11.12 and 2.11.21; Professor K. M. King, Ontario Agricultural College for figs. 4.1.4 and 4.1.5; R. P. Matthews of Portsmouth College of Technology for part of Table 3.1(i).2.

Finally, the following thanks are also due:

Mr A. Burn, Mr R. Smith, Mrs B. Human, and Miss L. Thorne of the Drawing Office, Department of Geography, Southampton University, for drawing the figures for Chapters 1.1., 3.1(i), 4.1 and 11.1; Dr R. M. Holmes, Inland Waterways Branch, Calgary, for valuable comments on Chapter 4.1; Dr Mark Meier, U.S. Geological Survey for material used in Chapter 8.1; Dr Gunnar Østrem, Glaciology Section, Vassdragsvesenet for material used in Chapter 8.1; Mr M. Young, Miss R. King, and Mr M. J. Ampleford of the Drawing Office, Department of Geography, Cambridge University, for drawing figures for Chapter 2.11; Dr K. A. Edwards and Members of the Institute of Hydrology, Wallingford, for their kind assistance with Chapters 3.1(ii) and 9.1, Mr V. J. Nash for preparing the illustrations.

The Editor and Publishers would like to thank Mrs D. M. Beckinsale for her painstaking and authoritative preparation of the Index, which has contributed greatly to the value of this volume.

Introduction

R. J. CHORLEY and R. W. KATES

Department of Geography, Cambridge University and Graduate School of Geography, Clark University

Who would not choose to follow the sound of running waters? Its attraction for the normal man is of a natural sympathetic sort. For man is water's child, nine-tenths of our body consists of it, and at a certain stage the foetus possesses gills. For my part I freely admit that the sight of water in whatever form or shape is my most lively and immediate form of natural enjoyment: yes, I would even say that only in contemplation of it do I achieve true self-forgetfulness and feel my own limited individuality merge into the universal.

(Thomas Mann: *Man and his Dog*)

1. 'Physical' and 'human' geography

Perhaps it is of the nature of scholarship that all scholars should think themselves to be living at a time of intellectual revolution. Judged on the basis of the references which they have cited (Stoddart, 1967, pp. 12–13), geographers have long had the impression that they were the immediate heirs of a surge of worthwhile and quotable research. There is good reason to suppose, however, that geography has just passed through a major revolution (Burton, 1963), one of the features of which has been profoundly to affect the traditional relationships between 'physical' and 'human' geography.

Ever since the end of the Second World War drastic changes have been going on in those disciplines which compose physical geography. This has been especially apparent in geomorphology (Chorley, 1965a), where these changes have had the general effect of focusing attention on the relationships between process and form, as distinct from the development of landforms through time. In the early 1950s geomorphologists, especially in Britain, were able to look patronizingly at the social and economic branches of geography and dismiss them as non-scientific, poorly organized, slowly developing, starved of research facilities, dealing with subject matter not amenable to precise statement, and denied the powerful tool of experimentation (Wooldridge and East, 1951, pp. 39–40). It is true that by this time most geographers had long rejected the dictum that physical geography 'controlled' human geography, but most orthodox practitioners at least paid lip service to the idea that there was a physical *basis* to the subject. This view was retained even though traditional geomorphology had little or nothing to contribute to the increasingly urban and industrial preoccupations of human geographers (Chorley, 1965b, p. 35), and its place in the subject as a whole was maintained either as a conditioned reflex or as increasingly embarrassing grafts on to new geographical shoots. American geographers, who had largely abandoned geomorphology to the geologists even before the war, tended to look more to climatology for their physical basis. However, despite the important researches of Thornthwaite and of more recent work exemplified by that of Curry [1952] and Hewes [1965], the proportion of articles relating to weather and climate appearing in major American geographical journals fell more or less steadily from some 37% in 1916 to less than 5% in 1967 (Sewell, Kates, and Phillips, 1968). Even in the middle of the last decade Leighly (1955, p. 317) was drawing attention to the paradox that instructors in physical geography might be required to teach material quite unrelated to their normal objects of research.

The problems of the relationships between physical and human geography facing Leighly were small, however, compared with those which confront us today. Little more than a decade has been sufficient to transform the leading edge of human geography into a 'scientific subject', equipped with all the quantitative and statistical tools the possession of which had previously given some physical geographers such feelings of superiority. Today human geography is not directed towards some unique areally-demarcated assemblage of information which can be viewed either as a mystical *gestalt* expressive of some 'regional personality' or simply as half-digested trivia, depending on one's viewpoint. In contrast, most of the more attractive current work in human geography is aimed at more limited and intellectually viable syntheses of the pattern of human activity over space possessing physical inhomogeneities, leading to the disentangling of universal generalizations from local 'noise' (Haggett, 1965). Today it is human geography which seems to be moving ahead faster, to have the more stimulating intellectual challenges, and to be directing the more imaginative quantitative techniques to their solution.

One immediate result of this revolution has been the demonstration, if this were further needed, that the whole of geomorphology and climatology is not coincident with physical geography, and that the professional aims of the former are quite distinct from those of the latter. This drawing apart of traditional physical and human geography has permitted their needs and distinctions, which had previously been obscure, to emerge more clearly. Perhaps the distinctions may have become too stark, as evidenced by current geographical preoccupations with a rootless regional science and with socio-economic games played out on featureless plains or within the urban sprawl. Perhaps this is what the future holds for geography, but it is clear that without some dialogue between man and the physical environment within a spatial context geography will cease to exist as a discipline.

There is no doubt that the major branches of what was previously called physical geography can exist, and in some cases already are existing, under the umbrella of the earth sciences, quite happily outside geography, and that they are probably the better for it. It is also possible that this will be better for geography in the long run, despite the relevance to it of many of the data and certain of the techniques and philosophical attitudes of the earth sciences. In their pl a more meaningful and relevant physical geography may emerge as the produ of a new generation of physical geographers who are willing and able to face up to the contemporary needs of the whole subject, and who are prepared to concentrate on the areas of physical reality which are especially relevant to the modern man-oriented geography. It is in the extinction of the traditional division between physical and human geography that new types of collaborative synthesis can arise. Such collaborations will undoubtedly come about in a number of ways, the existence of some of which is already a reality. One way is to take a philosophical attitude implied by an integrated body of techniques or models (commonly spatially oriented) and demonstrate their analogous application to both human and physical phenomena (Woldenberg and Berry, 1967; Haggett and Chorley, In press). Another way is to assume that the stuff of the physical world with which geographers are concerned are its resources – resources in the widest sense; not just coal and iron, but water, ease of movement, and even available space itself. In one sense the present volume represents both these approaches to integration by its concentration on the physical resource of water in all its spatial and temporal inequalities of occurrence, and by its conceptualization of the many systems subsumed under the hydrological cycle (Kates, 1967). In the development of water as a focus of geographical interest the evolution of a human-oriented physical geography and an environmentally sensitive human geography closely related to resource management is well under way.

2. Water as a focus of geographical interest

Water, Earth, and Man, both in organization and content, reflects the foregoing attitudes by illustrating the advantages inherent in adopting a unified view of the earth and social sciences. The theme of this book is that the study of water provides a logical link between an understanding of physical and social environments. Each chapter develops this theme by proceeding from the many aspects of water occurrence to a deeper understanding of natural environments and their fusion with the activities of man in society. In this way water is viewed as a highly variable and mobile resource in the widest sense. Not only is it a commodity which is directly used by man but it is often the mainspring for extensive economic development, commonly an essential element in man's aesthetic experience, and always a major formative factor of the physical and biological environment which provides the stage for his activities. The reader of this volume is thus confronted by one of the great systems of the natural world, the hydrologic cycle, following water through its myriad paths and assessing its impact on earth and man. The hydrologic cycle is a great natural system, but it should become apparent that it is increasingly a technological and social system as well. It has been estimated that 10% of the national wealth of the United States is found in capital structures designed to alter the hydrologic cycle: to collect, divert, and store about a quarter of the available surface water, distribute it where needed, cleanse it, carry it away, and return it to the natural system. The technical structures are omnipresent: dams, reservoirs, aqueducts, canals, tanks, and sewers, and they become increasingly sophisticated in the form of reclamation plants, cooling towers, or nuclear desalinization plants. The social and political system is also pervasive and equally complex, when one reflects on the number of major decision makers involved in the allocation and use of the water resources. White has estimated that for the United States the major decision makers involved in the allocation and use of water include at least 3,700,000 farmers, and the managers of 8,700 irrigation districts, 8,400 drainage districts, 1,600 hydroelectric power plants, 18,100 municipal water-supply systems, 7,700 industrial water-supply systems, 11,400 municipal sewer systems, and 6,600 industrial-waste disposal systems.

This coming together of natural potential and of human need and aspiration provides a unique focus for geographic study. In no other major area of geographic concern has there been such a coalescence of physical and human geography, nor has there developed a dialogue comparable to that which exists between geographers and the many disciplines interested in water. How these events developed is somewhat speculative. First, there is the hydrologic cycle itself, a natural manifestation of great pervasiveness, power, and beauty, that transcends man's territorial and intellectual boundaries. Equally important is that in the human use of water there is clear acknowledgement of man's dependence on environment. This theme, developed by many great teachers and scholars, (e.g. Ackerman, Barrows, Brunhes, Davis, Gilbert, Lewis, Lvovich, Marts, Powell, Thornthwaite, Tricart, and White), is still an important geographic concern, despite the counter trends previously described. Finally, there is no gainsaying the universal appeal of water itself, arising partly from necessity, but also from myth, symbol, and even primitive instinct.

The emergence of water as a field of study has been paralleled in other fields. In the application of this knowledge to water-resource development, a growing consensus emerges as to what constitutes a proper assessment of such development: the estimation of physical potential, the determination of technical and economic feasibility, and the evaluation of social desirability. For each of these there exists a body of standard techniques, new methods of analysis still undergoing development, and a roster of difficult and unsolved problems. Geographers have made varying contributions to these questions, and White reviewed them in 1963. Five years later, what appear to be the major geographic concerns in each area?

Under the heading of resource estimates, White cites two types of estimates of physical potential with particular geographic significance. The first is 'the generalized knowledge of distributions of major resources . . . directly relevant to engineering or social design'. While specific detailed work, he suggests, may be in the province of the pedologist, geologist, or hydrologist, there is urgent need for integrative measures of land and water potential capable of being applied broadly over large areas. The need for such measures has not diminished, but rather would seem enhanced by developments in aerial and satellite reconnaissance that provide new tools of observation, and by the widespread use of computers that provide new capability for data storage and analysis. In the developing world the need is for low-cost appraisal specific to region or project.

A second sort of estimate of potential that calls upon the skills of both the physical and human geographer is to illuminate what White calls 'the problem of the contrast between perception of environment by scientists . . . (and) others who make practical decisions in managing resources of land and water'. These studies of environmental perception have grown rapidly in number, method, and content. They suggest generally that the ways in which water and land resources receive technical appraisal rarely coincide with the appraisals of resource users. This contrast in perception is reflected in turn by the divergence between the planners' or technicians' expectation for development and the actual course of development. There are many concrete examples: the increase in flood damages despite flood-control investment, the almost universal lag in the use of available irrigation water, the widespread rejection of methods of soil conservation and erosion control, and the waves of invasion and retreat into the margins of the arid lands. Thus a geography that seeks to characterize environment as its inhabitants see it provides valued insight for the understanding of resource use.

In 1963 White differentiated between studies of the technology of water management and studies of economic efficiency. Today one can suggest that, increasingly, technical and economic feasibility are seen as related questions. The distinction between these areas, one seen as the province of the engineer and hydrologist, the other as belonging to the economist and economic geographer, is disappearing, encouraged by the impressive results of programmes of collaborative teaching and research between engineering and economics (e.g. at Stanford and Harvard Universities). In this view, the choice of technology and of scale is seen as a problem of cost. The choice of dam site, construction material, and height depends on a comparison of the incremental costs and of the incremental benefits arising from a range of sites, materials, and heights. This decision can be simultaneously related through systems analysis to the potential outputs of the water-resource system.

The methodology for making such determinations has probably outrun our understandings of the actual relationships. The costs and benefits of certain technologies are not always apparent, nor are all the technologies yet known. Geographic research on a broadened range of resource use and specific inquiry into the spatial and ecological linkages (with ensuing costs) of various technologies appears to be required. Indeed, as the new technologies of weather forecasting and modification, desalinization, and cross-basin transport of water and power expand, the need for such study takes on a special urgency.

Finally, there appears to be a growing recognition that much of what may be socially important in assessing the desirability of water-resource development will escape our present techniques of feasibility analysis for much time to come. The need for a wider basis of choice to account for the social desirability of water-resource development persists and deepens as the number of water-related values increase and the means for achieving them multiply. A framework for assessing social desirability still needs devising, but it could be hastened by careful assessment of what actually follows water-resource development. There is much to be learned from the extensive developments planned or already constructed. However, studies such as Wolman's [1967] attempt to measure the impact of dam construction on downstream river morphology or the concerted effort to assess the biological and social changes induced by the man-made lakes in Africa are few and far between. Studies built on the tradition of geographic field research but employing a rigorous research design over an extended period of observation are required. Geographers, freed from the traditional distinction between human and physical geography and with their special sensitivity towards water, earth, and man, have in these both opportunity and challenge.

REFERENCES

ACKERMAN, E. A. [1965], The general relation of technology change to efficiency in water development and water management; In Burton I. and Kates, R., Editors, *Readings in Resource Management and Conservation* (Chicago), pp. 450–67.

BURTON, I. [1963], The quantitative revolution and theoretical geography; *The Canadian Geographer*, **7**, 151–62.

BURTON, I. and KATES, R. [1964], The perception of natural hazards in resource management; *Natural Resources Journal*, **3**, 412–41.

CHORLEY, R. J. [1965a], The application of quantitative methods to geomorphology; In Chorley, R. J. and Haggett, P., Editors, *Frontiers in Geographical Teaching* (Methuen, London), pp. 147–63.

CHORLEY, R. J. [1965b], A re-evaluation of the geomorphic system of W. M. Davis; In Chorley, R. J. and Haggett, P., Editors, *Frontiers in Geographical Teaching* (Methuen, London), pp. 21–38.

CURRY, L. [1952], Climate and economic life: A new approach with examples from the United States; *Geographical Review*, **42**, 367–83.

HAGGETT, P. [1965], *Locational Analysis in Human Geography* (Arnold, London), 339 p.

HAGGETT, P. and CHORLEY, R. J. [1969], *Network Models in Geography* (Arnold, London).

HEWES, L. [1965], Causes of wheat failure in the dry farming region, Central Great Plains, 1939–57; *Economic Geography*, **41**, 313–30.

HUFSCHMIDT, M. [1965], The methodology of water-resource system design; In Burton, I. and Kates, R., Editors, *Readings in Resource Management and Conservation* (Chicago), pp. 558–70.

KATES, R. W. [1967], Links between Physical and Human geography; In *Introductory Geography: Viewpoints and Themes* (Washington), pp. 23–31.

LEIGHLY, J. [1955], What has happened to physical geography?; *Annals of the Association of American Geographers*, **45**, 309–18.

SEWELL, W. R. D., Editor [1966], Human Dimensions of Weather Modification; *University of Chicago, Department of Geography, Research Paper* **105**, 423 p.

SEWELL, W. R. D., KATES, R. W., and PHILLIPS, L. E. [1968], Human response to weather and climate; *Geographical Review*, **58**, 262–80.

STODDART, D. R. [1967], Growth and structure of geography; *Transactions of the Institute of British Geographers*, No. 41, 1–19.

WHITE, G. F. [1963], Contribution of geographical analysis to river basin developme *Geographical Journal*, **129**, 412–36.

WHITE, G. F. [1968], *Strategies of American Water Management* (Ann Arbor).

WOLDENBERG, M. J. and BERRY, B. J. L. [1967], Rivers and central places: Analogous systems?; *Journal of Regional Science*, **7** (2), 129–39.

WOLMAN, M. G. [1967], Two problems involving river channel changes and background observations; In Garrison, W. L. and Marble, D. F., Editors, *Quantitative Geography: Part II Physical and Cartographic Topics* (Northwestern University), pp. 67–107.

WOOLDRIDGE, S. W. and EAST, W. G. [1951], *The Spirit and Purpose of Geography* (Hutchinson, London), 176 p.

orld Hydrological Cycle

R. G. BARRY
Institute of Arctic and Alpine Research, University of Colorado

1. Global water and the components of the hydrological cycle

We begin our consideration of water in the global context with some figures to illustrate the storage capacity of the earth–atmosphere system. The oceans, with a mean depth of 3·8 km and covering 71% of the earth's surface, hold 97% of *all* the earth's water ($1{\cdot}31 \times 10^{24}$ cm³). 75% of the total *fresh* water is locked up in glaciers and ice sheets, while almost all of the remainder is ground water. It is an astonishing fact that at any instant rivers and lakes hold only 0·33% of all fresh water and the atmosphere a mere 0·035% (about 12×10^{18} cm³).

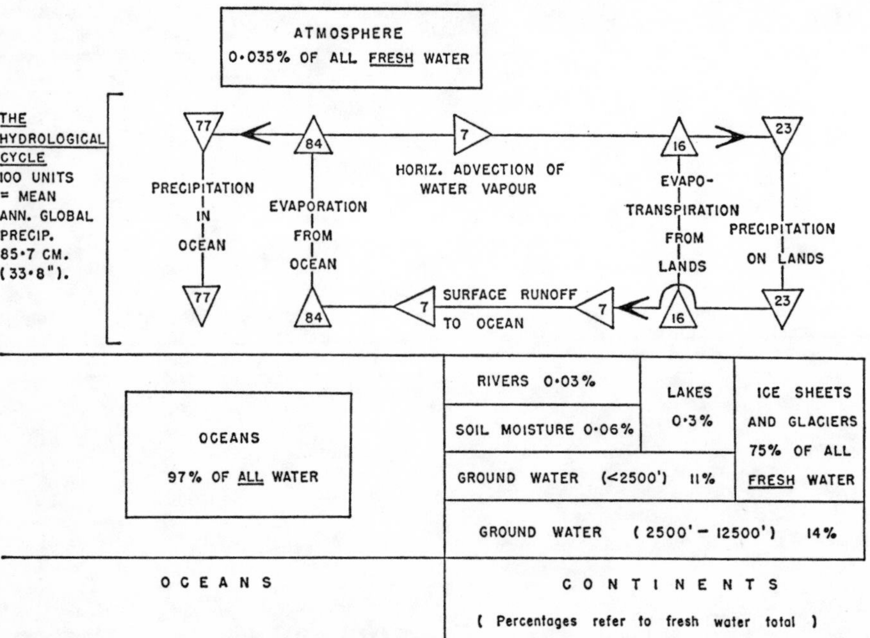

Fig. 1.1.1 The global hydrological cycle and water storage.

The exchanges in the cycle are referred to 100 units which equal the mean annual global precipitation of 85·7 cm (33·8 in.). The storage figures for atmospheric and continental water are percentages of all *fresh* water. The saline oceans make up 97% of *all* water (From More, 1967).

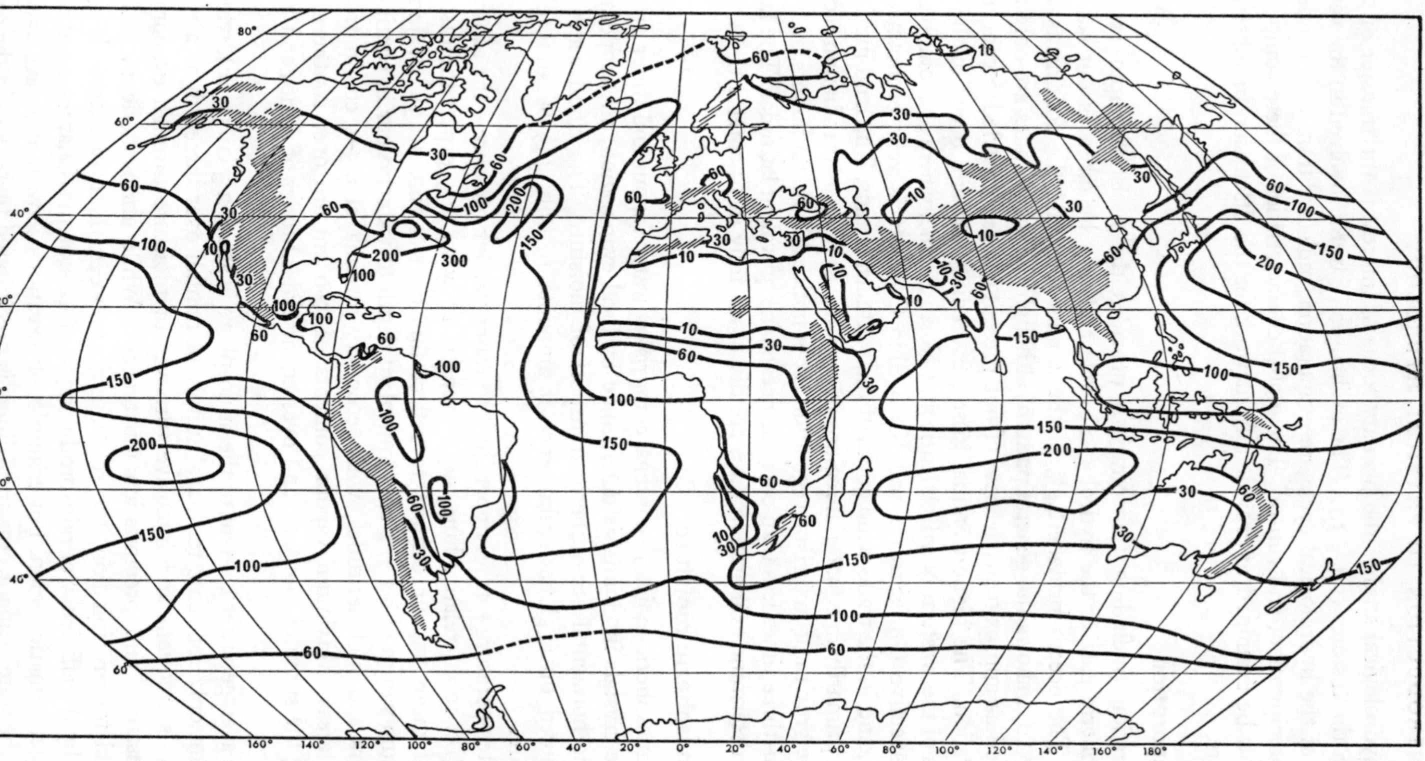

Fig. 1.1.2 Annual evaporation in cm (After Budyko *et al.*, 1962).

In hydrological studies the primary focus of interest is the transfer of water *between* these stores (fig. 1.1.1). The exchanges of water involved in the various stages of the hydrological cycle are evaporation, moisture transport, condensation, precipitation, and run-off. The global characteristics of these components will now be examined to provide a framework for the discussion in subsequent chapters.

2. Evaporation

Evaporation (including transpiration) provides the moisture input into the atmospheric part of the hydrological cycle and may be taken as our starting-point. The oceans provide 84% of the annual total and the continents 16%. Figure 1.1.2 shows the general pattern, although the magnitudes are only to be regarded as approximate in view of our present limited knowledge concerning evaporation. The highest annual losses, exceeding 200 cm, occur in the subtropics of the western North Atlantic and North Pacific, where evaporation over the respective Gulf Stream and Kuro Shio Currents is very pronounced in winter, and in the trade-wind zones of the southern oceans. The land maximum occurs primarily in equatorial regions in response to high solar radiation receipts and the growth of luxuriant vegetation. It is noticeable that amounts over land are two–three times less than over the oceans in equivalent latitudes. The factors which determine evaporation rates are discussed fully in Chapter 4.1.

3. Atmospheric moisture

The atmospheric moisture content, comprising water vapour and water droplets and ice crystals in clouds, is determined by local evaporation, air temperature, and the horizontal atmospheric transport of moisture. The cloud water may be ignored on a global scale, since it amounts to only 4% of atmospheric moisture.

Air temperature sets an upper limit to water-vapour pressure – the saturation value (i.e. 100% relative humidity) – consequently we may expect the distribution of mean vapour content to reflect this control (fig. 1.1.3). In January minimum values of 0·1–0·2 cm (equivalent depth of water) occur in continental interiors and high latitudes, with secondary minima of 0·5–1 cm in tropical desert areas. Maximum vapour contents of 5–6 cm are over southern Asia during the summer monsoon and over equatorial latitudes of Africa and South America.

The average water content of the atmosphere is about 2·5 cm (1 in.), which is sufficient only for some ten days' supply of rainfall over the earth as a whole. Clearly, a frequent and intensive turnover of moisture through evaporation, condensation, and precipitation must occur. While atmospheric moisture is essential for precipitation, the relationship between these two items is determined by the efficiency of rain-producing weather systems (Chapter 3.1.2) in any particular climatic region. For example, observations show that on average only 5% of the water vapour crossing Illinois is precipitated there, and in the case of the Mississippi basin only about 20%.

4. Precipitation

The major types of precipitation are drizzle, rain, snow, and hail, although dew, fog drip, hoar frost, and rime may also make significant contributions to the total (see Chapter 3.1). The distribution of summer and winter precipitation is shown in fig. 1.1.4. The least reliable parts of the maps are the oceans, especially in the southern hemisphere. Analysis of precipitation frequency at North Atlantic weather ships suggests that previous estimates of annual totals in the north-western sector are 20–50 cm too low, while around 45° N, 15° W they are 40–50 cm too high.

The patterns reflect many complex weather factors and geographical influences, such as topography and the land–sea distribution, but the most significant features are:

1. The 'equatorial' maximum, which is displaced into the northern hemisphere. This is related primarily to the converging trade-wind systems and monsoon regimes of the summer hemisphere, particularly in southern Asia and West Africa. Annual totals over large areas are of the order of 200–250 cm (80–100 in.) or more.
2. The west coast maxima of middle latitudes associated with the belt of travelling disturbances in the westerlies. The precipitation in these areas has a very high degree of reliability.
3. The dry areas of the subtropical high-pressure cells, which include not only many of the world's major deserts but also vast oceanic expanses. In the northern hemisphere the remoteness of the continental interiors extends these dry conditions into middle latitudes. In addition to very low average annual totals, often less than 15 cm (6 in.), these regions are subject to considerable year-to-year variability.
4. Low precipitation in high latitudes and in winter over the continental interiors of the northern hemisphere. This reflects the low vapour contents of the extremely cold air.

Types of seasonal regime and other precipitation characteristics are examined in Chapter 3.1.

5. Water circulation in the atmosphere

The previous sections have been concerned only with the static aspects of moisture transfer at the surface and storage in the air, but the atmospheric transport of moisture is an important aspect of climatic differentiation over the earth. Comparison of annual average precipitation and evaporation totals for latitude zones shows that in low and middle latitudes $P > E$, whereas in the subtropics $P < E$ (fig. 1.1.5). These regional imbalances are maintained by net moisture transport into (convergence) and out of (divergence) the respective zones (ΔD, where divergence is positive).

$$E - P = \Delta \mathrm{D} \qquad (1)$$

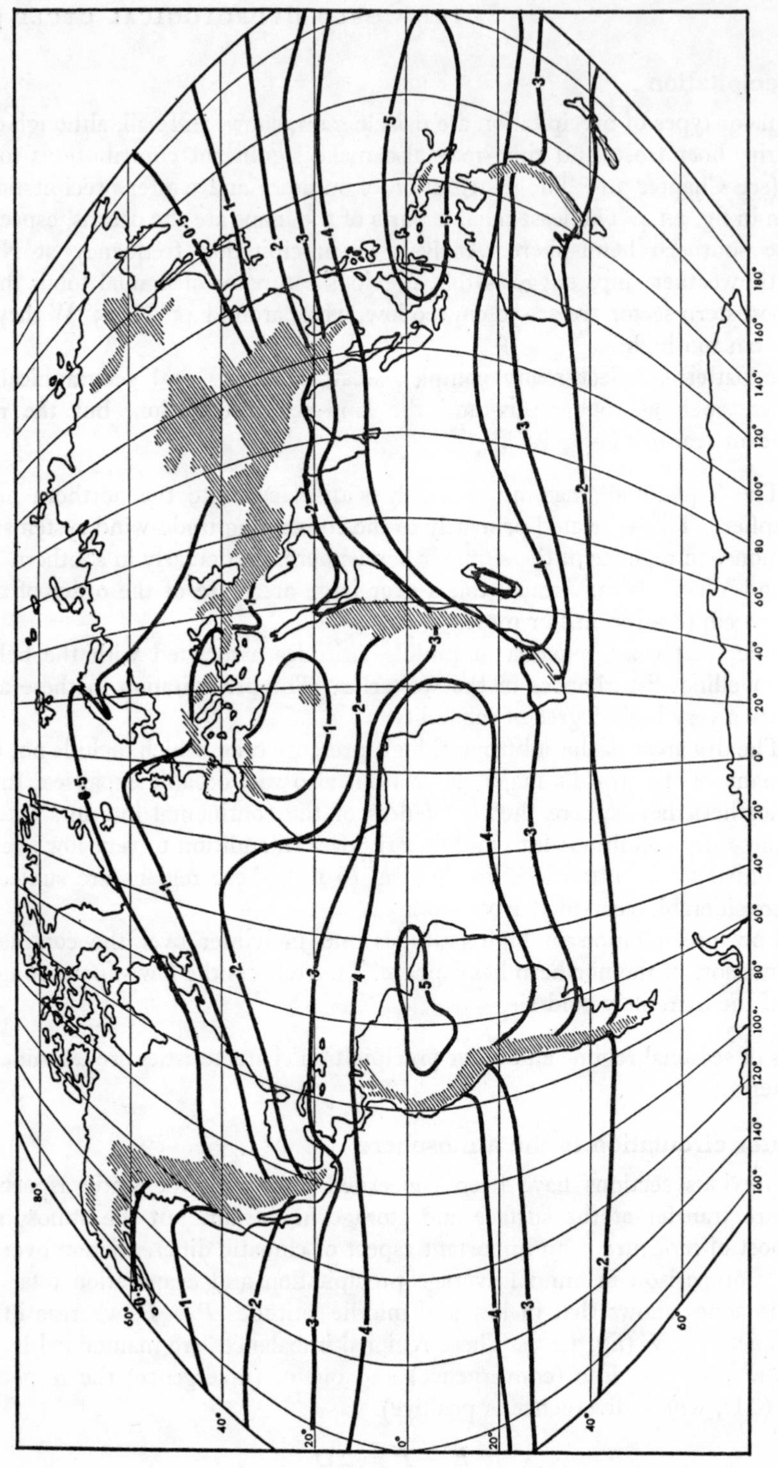

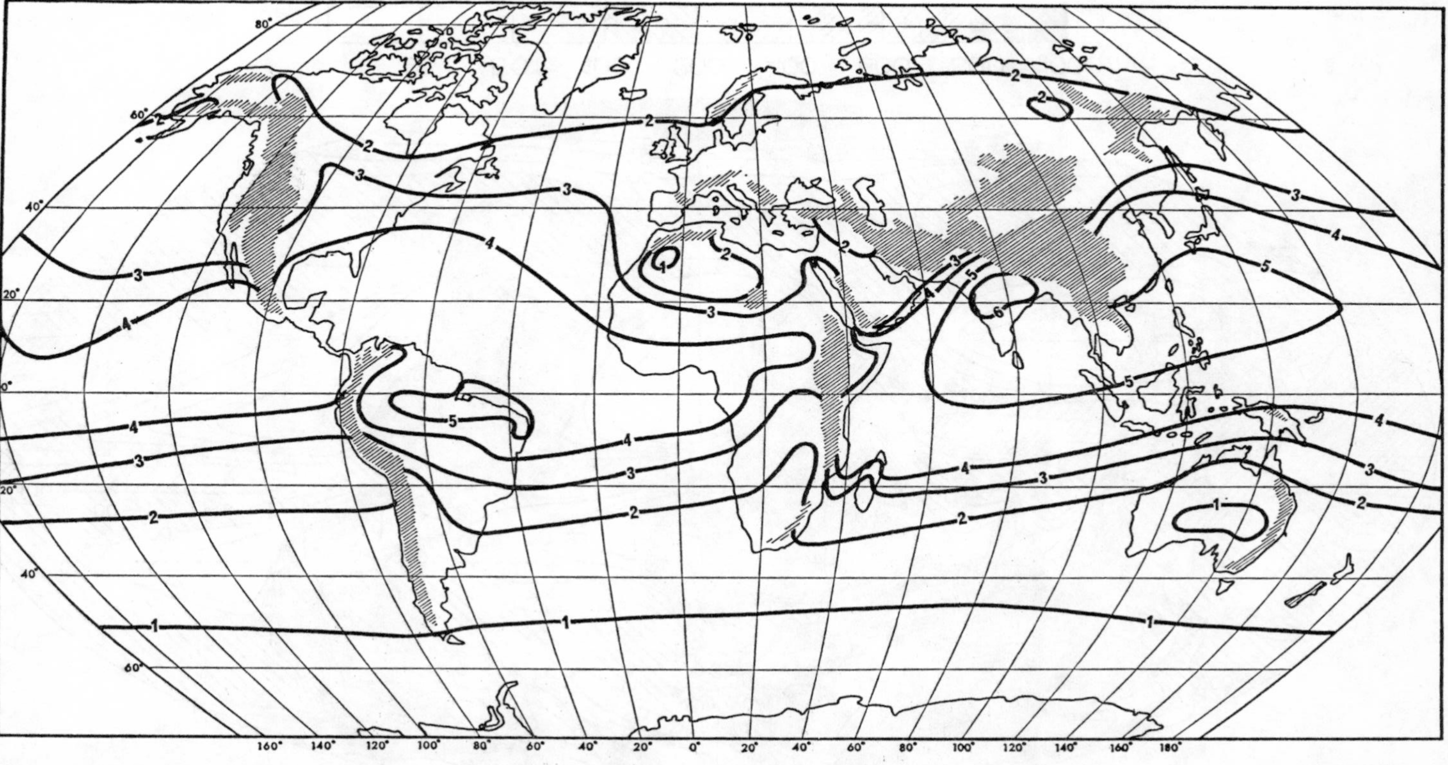

Fig. 1.1.3 Mean atmospheric water vapour content in January (*above*) and July (*below*) 1951–5, in cm of precipitable water (After Bannon and Steele, 1960). (Crown Copyright Reserved).

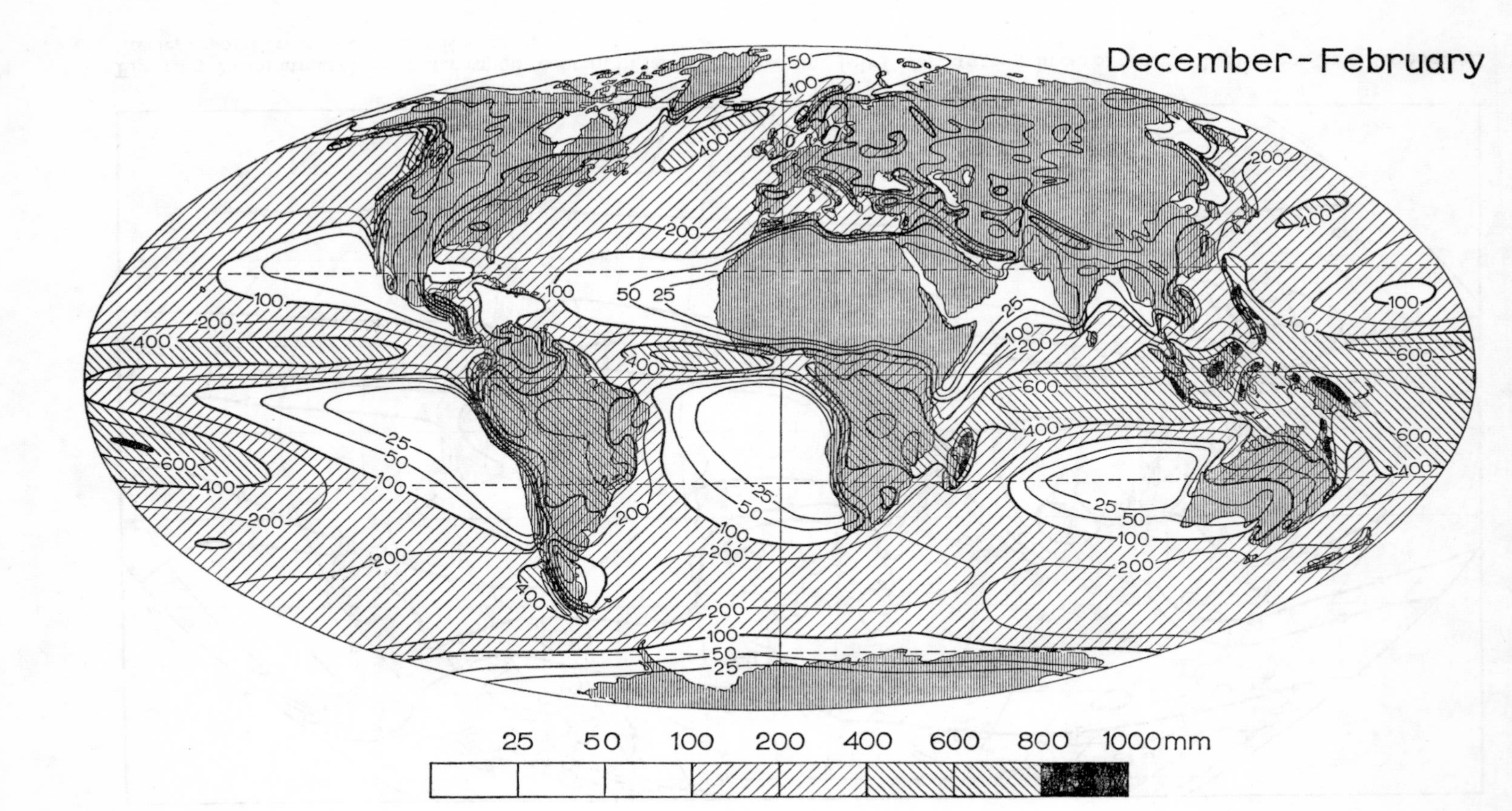
December - February
25
50
100
200
400
600
800
1000mm

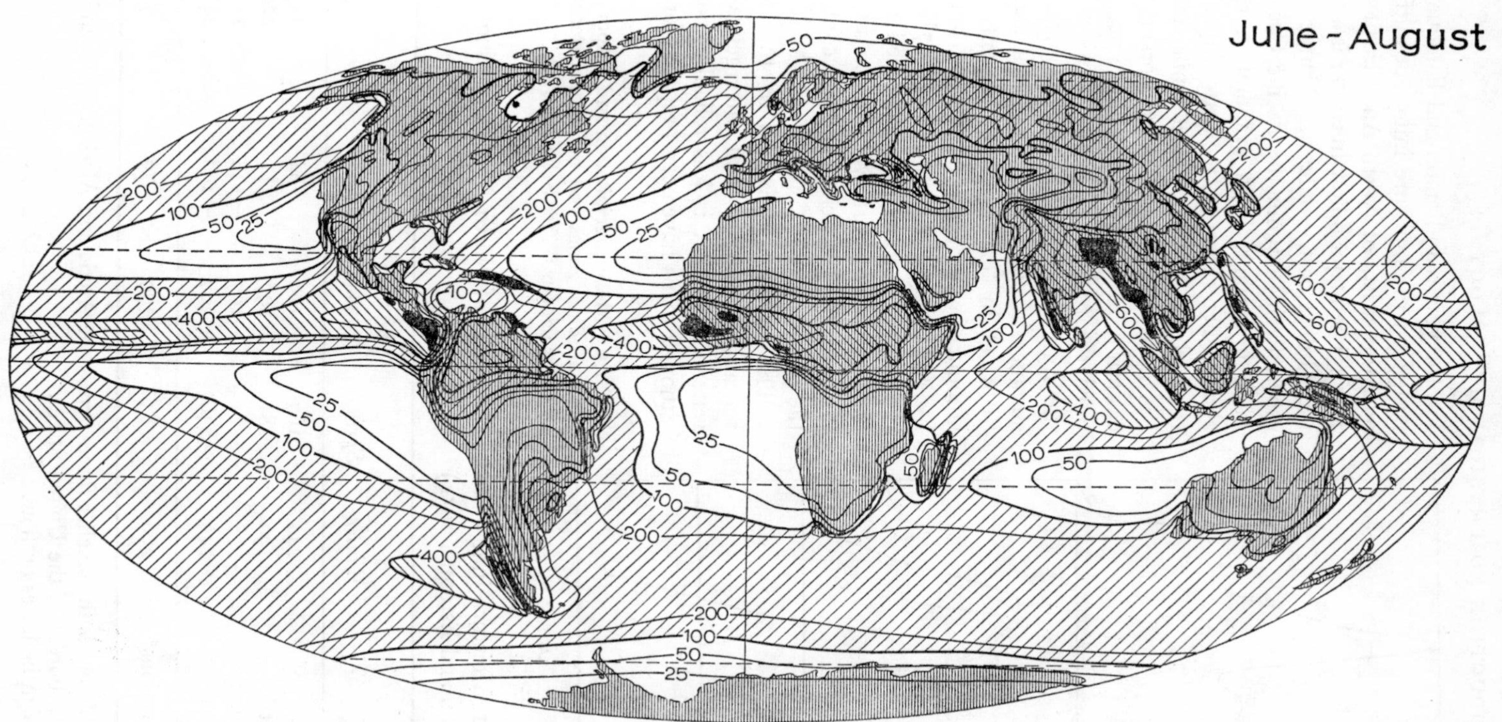

Fig. 1.1.4 Global precipitation for December–February (*above*) and June–August (*below*) in mm (From Möller, 1951).

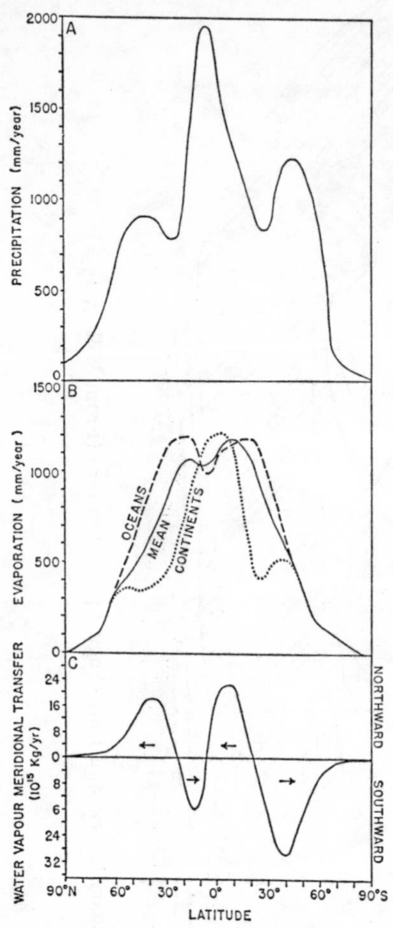

For the year 1949 Benton and Estoque investigated the moisture inflow or outflow over the coasts of North America (Table 1.1.1). The major inflow in winter is across the Gulf Coast, while in summer and especially in autumn the Pacific Coast is more important. High evaporation rates allow a net *export* of moisture from the continent in summer, and a similar result has been obtained by other workers for eastern Asia. These findings have necessitated the revision of earlier views about the role of oceanic moisture sources in summer in both areas. The total annual 'balance' in Table 1.1.1 represents the amount of water which must be discharged by the rivers. Rasmusson [1967] has recently carried out a more detailed study for vapour transport over North America during 1961–3 which essentially confirms the previous results.

The spatial distribution of the horizontal convergence and divergence of water-vapour flux is less reliably known, although maps have been prepared for the

Fig. 1.1.5 Mean precipitation and evaporation for latitudinal zones and meridional transport of water vapour (After Sellers, 1965).

TABLE 1.1.1 Moisture inflow across the coasts of North America during 1949 (after Benton and Estoque, 1954)

			Units: 10^6 kg/sec		
	Winter	*Spring*	*Summer*	*Autumn*	*Year*
Gulf Coast	244	167	168	84	157
Pacific Coast	190	181	197	311	220
Labrador Coast	−78	−64	−167	−148	−114
Atlantic Coast	−249	−248	−307	−190	−248
All coastal sections*	206	83	−79	79	72

* This table omits the smaller fluxes across the Arctic and Alaskan coasts and the south-western border of the United States.

A minus sign indicates outflow.

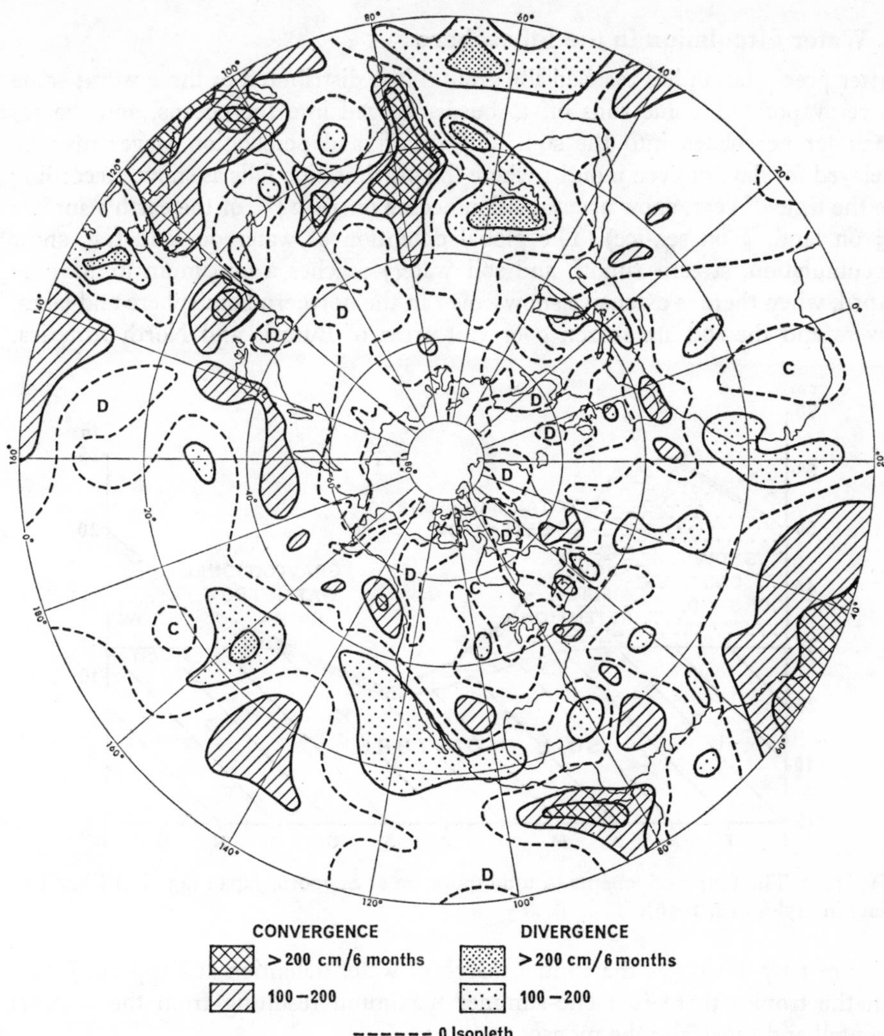

Fig. 1.1.6 The horizontal divergence of water-vapour transport for the six summer months, 1958 (After Peixoto and Crisi, 1965).
Divergence is positive, convergence negative. The units are cm of water/6 months.

northern hemisphere for 1950 and 1958, as well as for continental and other areas. Figure 1.1.6 shows a map for the six summer months of 1958 prepared by Peixoto and Crisi. In studying such maps it is essential to remember that only the *balance* of precipitation and evaporation is shown. Prominent features are the divergent zones, i.e. $P - E < 0$, of the oceanic subtropical high-pressure systems and the convergent areas of monsoon regime in India and Malaysia. The large divergence, implying a moisture source, east of the Persian Gulf is probably a spurious result due to the shortcomings in the available data.

6. Water circulation in the lithosphere

After precipitation has reached the ground it is distributed in three ways; some is re-evaporated, some runs off to be discharged into the oceans, and the remainder percolates into the soil. The runoff and percolation stages may be delayed for days or even months where the precipitation falls as snow. According to the time of year, snow accumulation occurs on 14–24% of the earth's surface ($\frac{2}{3}$ on land, $\frac{1}{3}$ on sea-ice). The global detention of water on land (i.e. snow accumulation, stream runoff, and soil water) reaches a maximum in March–April, when there is extensive snow-cover in the northern hemisphere and lakes, rivers and the soil are frozen over vast areas of Eurasia and North America.

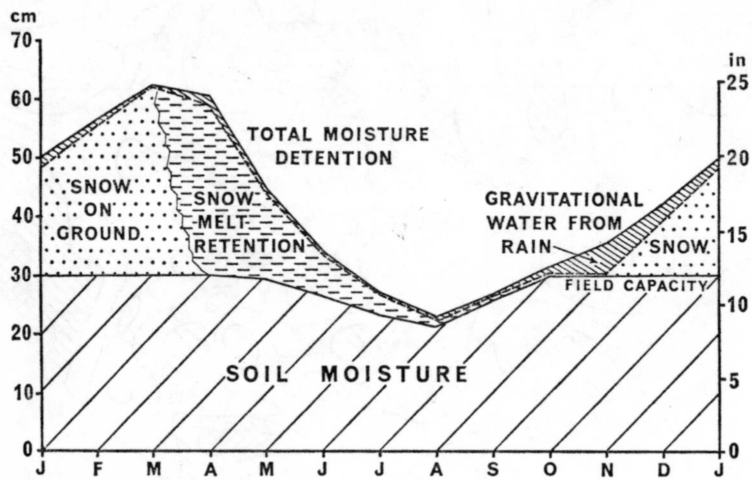

Fig. 1.1.7 The four components of total moisture at Sapporo, Japan (43° N.) (Based on data in Hylckama, 1956).

Figure 1.1.7 illustrates the annual march of water detention at Sapporo, Japan. In the tropics there is a late-summer maximum resulting from the summer rainfall and especially the monsoons.

The seasonal variation in global detention is matched by an inverse pattern of storage in the oceans. In October the seas are estimated to hold $7{\cdot}5 \times 10^{18}$ cm^3 more water than in March, although this is equivalent to a sea-level change of only 1–2 cm.

In the above situation the residence time of water on the land is comparatively short, approximately 10^0–10^2 days. In the case of glaciers and ice-caps, however, the storage time is of the order of years. In the extreme case of central Antarctica it is estimated by Shumskiy and his co-workers [1964] that the maximum storage time of ice is about 200,000 years.

Ground water is a similarly stable component of the hydrological cycle. Most ground water represents precipitation which has percolated through the soil layers into the zone of saturation, where all interstices are water-filled. Water of this origin is termed *meteoric.* Minor sources of water are located in the earth's crust. They are *connate* water, representing water trapped during the formation of sedimentary rocks, and *juvenile* water. The latter, earlier considered to be reaching the surface for the first time in connection with volcanic activity, is now thought to be mainly connate. In many arid areas with internal drainage the major source of ground water is seepage from stream runoff and lakes. Near the water-table the cycling-time of water is a year or less, but in deep aquifers it is of the order of thousands of years (see Chapter 1.11). A similar storage-time applies in the case of ground ice in the permafrost regions. In addition, there are deep-seated brines (connate water), which are effectively isolated from the hydrological cycle, where any circulation has a time-scale of geological epochs.

Ground water contributes on average approximately 30% of total runoff, although within different geographical zones this proportion varies considerably. Some illustrations based on calculations by L'vovich for river basins in the

TABLE 1.1.2 Run-off relationships for selected river basins in different geographical zones of the U.S.S.R. (after L'vovich, 1961)

Zone	*Precipitation* (cm)	*Evaporation* (cm)	*Runoff* / *Precipitation* (%)	*Surface Runoff* / *Total Runoff* (%)
Tundra	45	11	76	97
Taiga with permafrost	40	21	49	95
Taiga	50	16	69	73
Mixed forest	58	37	36	76
Mixed forest	61	43	30	60
Forest steppe	41	33	19	79
Steppe	35	32	9	87
Semi-desert	20	19	5	90

U.S.S.R. are shown in Table 1.1.2. The proportion of precipitation going into runoff decreases as the heat available for evaporation increases. The effect of frozen ground on ground-water discharge is pronounced in the tundra and Siberian sections of the taiga. In the drier areas precipitation is mainly evaporated after moistening the soil, and consequently the contribution of ground water to streamflow is negligible. The volume of surface runoff going into streamflow reaches a maximum in the tundra and taiga zones.

Average runoff from the land masses can be estimated from precipitation and evaporation data if allowance is made for areas of inland drainage. Such areas, predominantly in Asia, Africa, and Australia, account for 25% of stream runoff. For all continents the average annual runoff is 26·7 cm (10·5 in.), although amounts far exceed this in South America and the Malayan Archipelago. The latter region provides 12% of the total runoff from only 2% of the land surface

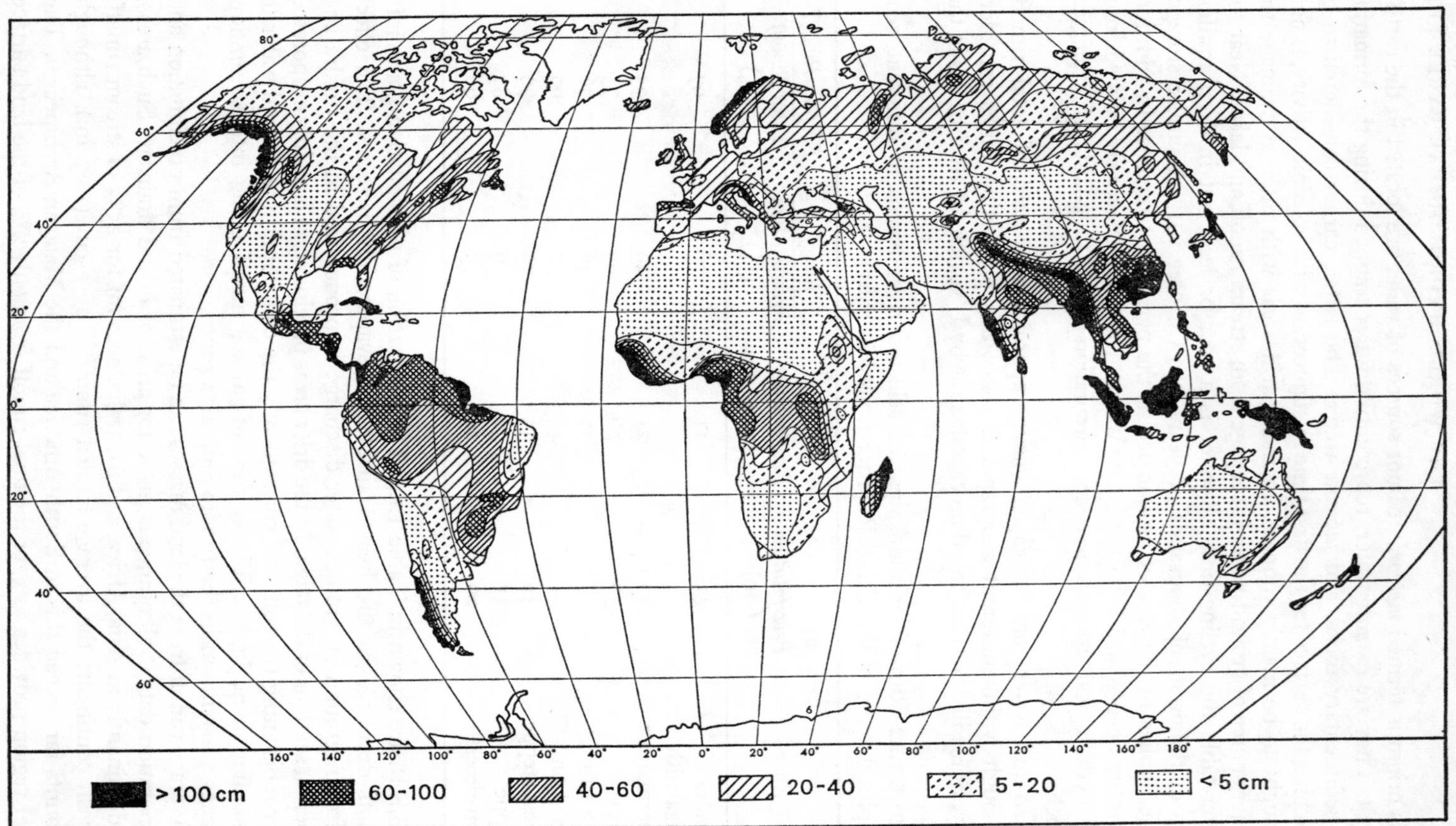

Fig. 1.1.8 Annual runoff in cm. Information is not available for areas left blank (After L'vovich, 1964).

(see Table 1.1.3). A more detailed picture of runoff is given in fig. 1.1.8. This is based on Soviet sources, which provide some of the most extensive information on the subject.

TABLE 1.1.3 Annual Runoff (after L'vovich)

	Land area		*% of total global runoff*
	10^6 *km*^{2}	10^6 *ml*^{2}	
Africa	29·8	11·5	17
Asia	42·2	16·3	20
Australia and New Zealand	8·0	3·1	2
Europe	9·6	3·7	7
Greenland	3·8	1·5	2
Malayan Archipelago	2·6	1·0	12
North America, Central America, and the Caribbean Lands	20·5	7·9	18
South America	17·9	6·9	22

7. The global water balance

Study of the global water balance has been made by Budyko and his co-workers in the Soviet Union (Table 1.1.4). The figures must obviously be regarded as first approximations, but they are quite adequate to convey the general picture. As far as individual oceans are concerned, their water balance involves not only precipitation (P), evaporation (E), and runoff (r) but also water exchange between oceans (ΔW).

$$P + r = E \pm \Delta W \qquad (2)$$

The magnitude of these components is shown in Table 1.1.4. For the continents, the water balance equation is

$$P = E + r \qquad (3)$$

and Table 1.1.5 summarizes the computations.

TABLE 1.1.4 Water balance of the Oceans (cm/year) (after Zubenok, in Budyko, 1956)

	Precipitation	*Runoff from adjoining land areas*	*Evaporation*	*Water exchange with other oceans*
Atlantic Ocean	78	20	104	−6
Arctic	24	23	12	35
Indian	101	7	138	−30
Pacific	121	6	114	13

TABLE 1.1.5 Water balance of the continents (cm/year)

	Precipitation	*Evaporation*	*Runoff*
Africa	67	51	16
Asia	61	39	22
Australia	47	41	6
Europe	60	36	24
North America	67	40	27
South America	135	86	49

Combining these data shows that for the earth as a whole precipitation and evaporation are of the order of 100 cm. The difference in the runoff figures in Table 1.1.6 is due, of course, to the respective surface areas of the oceans and continents. It is worth noting at this point that for each latitude zone around the globe the net convergence or divergence (equation (1)) is equivalent to the total runoff in that zone.

TABLE 1.1.6 The global water balance (cm/year) (after Budyko *et al.*, 1962)

	Precipitation	*Evaporation*	*Run-off*
Oceans	112	125	−13
Continents	72	41	31
Whole Earth	100*	100	0

* The computed value is actually 102 cm.

8. Global water circulation in relation to the energy budget

Water circulation in the atmosphere and oceans is intimately linked with the global energy budget. This is not the place for a detailed account, but it is appropriate to stress the interdependence of the moisture and energy budgets. Fuller details may be found in Sellers [1965, p. 100].

The annual input of solar radiation into the earth–atmosphere system and the net loss of terrestrial radiation produce a positive energy budget in low latitudes and a negative one in middle and higher latitudes. For annual averages the budget is balanced at about 35° latitude. Consequently, poleward heat transport is essential if the higher latitudes are not to become progressively colder and lower latitudes hotter. This transport occurs in three forms – atmospheric transport of sensible heat and of latent heat (i.e. water vapour which subsequently condenses), and the transport of warm water by ocean currents. The role of these three components is shown in fig. 1.1.9. Some 80% of the poleward heat transport takes place in the atmosphere. Sensible heat transport, by warm air masses, need not concern us here, although it accounts for the majority of the heat transport. The transport of latent heat reflects the pattern of the global wind belts on either side of the subtropical high-pressure zones, so that about 10° N and 10° S the flux is equatorward as a result of the transport of moisture into the equatorial low-pressure trough by the trade winds. Ocean currents, predominantly of course the Gulf Stream and Kuro Shio in the northern hemisphere and the Brazil Current and the poleward branches of the Equatorial Currents in the south-west Pacific and south-west Indian Ocean in the southern hemisphere, are most important about 35°–40° latitude.

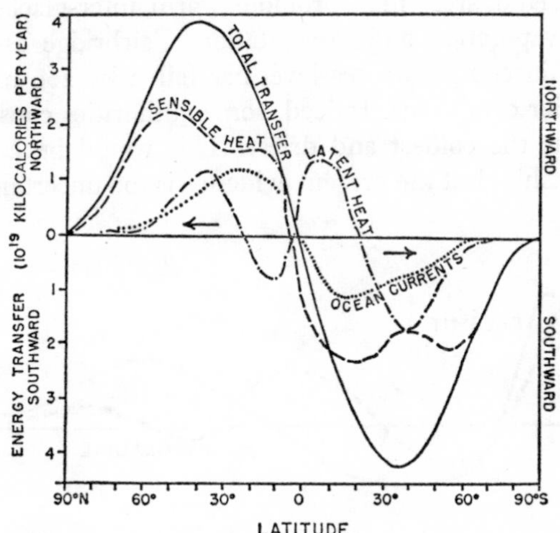

Fig. 1.1.9 Components of the poleward energy flux in the earth–atmosphere system (From Sellers, 1965).

9. Climatic change and the global water budget

The interdependence of the various components of the hydrological cycle makes it apparent that a change in any climatic parameter may have far-reaching repercussions. For instance, a 1% increase in evaporation from the tropical oceans would cool a 200-m layer by 3° C in fifty years (Malkus, 1962). Quantitative assessment of actual changes is obviously uncertain owing to observational limitations, but in the case of sea-level fluctuations, at least, the picture is now reasonably complete for late and post-glacial times.

It is estimated that at the Quaternary glacial maximum – the Illinoian, Riss, or Saale glacial – ice affected an area 3 × that of the present ice cover with 5 × the present mass. As a result of this long-term storage of water on the land, and the consequent diminution of run-off, a eustatic lowering of sea-level took place. Crary estimates the eustatic lowering as follows:

Glacial phase	*Eustatic lowering*	
Classical Wisconsin	105·5 m	348 ft
Early Wisconsin	114·5	378
Illinoian	137·4	453

Different calculations of ice volume by Novikov indicate a lowering of 159 m during the Illinoian maximum. The actual sea-level change in the glacierized areas was, of course, complicated by the isostatic depression of the continents due to their ice load.

One possible result of a falling sea-level is a decrease in evaporation and precipitation because a greater land area is exposed. Conversely, a sea-level rise, by reducing the land area, may promote warm inter-glacial conditions with higher average evaporation and precipitation. Fairbridge estimates that since the end of the Pliocene period sea-level has fallen by 200 m, and this should favour colder, drier conditions. Indeed, some authorities consider that the Last Glacial phase was the coldest and driest. This would be consistent with the suggested relationship, but the present evidence is too uncertain for confirmation

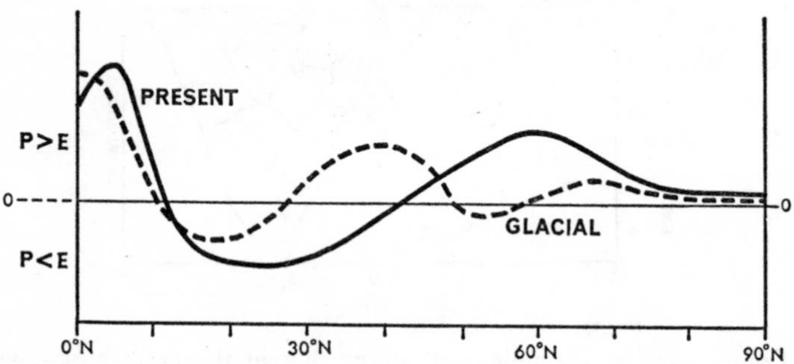

Fig. 1.1.10 The location of humid and arid zones in glacial times and the present day (From Flohn, 1953).

of this hypothesis. There is, in any case, little doubt that more important modifications of the precipitation–evaporation cycle took place due to the global cooling during glacial phases. In lower-middle latitudes evaporation, and consequently precipitation, was probably reduced by 20% at the full glacial stage, creating arid conditions in subtropical areas such as the Mediterranean. However, in the *early* stages of the glacial phases, before the oceans had begun to cool, the subtropics experienced 'pluvial' conditions (Butzer, 1957). In the south-west United States Pleistocene lake levels provide further evidence of subtropical pluvials. It is estimated that in order to maintain lakes at their maximum level, annual precipitation must have been 25 cm (10 in.) greater than now in north-central New Mexico, and 20 cm (8 in.) greater in east-central Nevada, with reduced evaporation due to cooler summers.

A model of the shift of the moisture budget zones under glacial conditions compared with the present has been outlined by Flohn [1953] as shown in fig. 1.1.10. Considerable work is required to refine such concepts and extend their application to the various phases of the glacial–interglacial cycle, and even when this has been satisfactorily achieved it will be necessary to establish appropriate correlations between climatic and hydrological parameters. The runoff during pluvial phases must have increased proportionately more in semi-arid than sub-humid climates (Schumm, 1965), but only tentative estimates of the magnitude of these changes can be made. Indeed, in some cases the sign of the change is still a matter of controversy. The implications of such changes are discussed in Chapter 11.II.

As climate started to ameliorate towards the end of the last glacial period the increased runoff from the melting ice-caps caused an eustatic rise in sea-level. This is considered to have begun about 18,000 years ago, when sea-level was some 105–120 m below the present level, and proceeded rapidly until about 6,000 years ago. Authorities disagree as to whether or not the level at that time exceeded present mean sea-level. One view is that it was approximately 5 m higher, while another school of thought considers that the post-glacial rise has been continuous, but diminished markedly over the last few thousand years. Whichever is correct, the relative stability during recent times indicates that the ice-caps of Antarctica and Greenland are more or less in equilibrium with climatic conditions. Direct measurements of sea-level since the end of the last century show a general rise of about 0·2 m/50 years up to 1940 and a 40% decrease in the rate of rise since then.

REFERENCES

BANNON, J. K. and STEELE, L. P. [1960], *Average Water Vapour Content of the Air; Geophysical Memoir No. 102* (H.M.S.O., London), 38 p.

BENTON, G. S. and ESTOQUE, M. A. [1954], Water-vapor transfer over the North American Continent; *Journal of Meteorology*, **11**, 462–77.

BUTZER, K. W. [1957], The recent climatic fluctuation in lower latitudes and the general circulation of the Pleistocene; *Geografiska Annaler*, **39**, 105–13.

BUDYKO, M. I. [1956], *The Heat Balance of the Earth's Surface* (Leningrad), translation by N. A. Stepanova (Washington), 1958, 259 p.

BUDYKO, M. I., *et al.* [1962], The heat balance of the surface of the earth; *Soviet Geography*, **3**, No. 5, 3–16.

CHOW, VEN TE, Editor [1964] *Handbook of Applied Hydrology* (New York).

DONN, W. L., FARRAND, W. R., and EWING, M. [1962], Pleistocene sea volumes and sea-level lowering; *Journal of Geology*, **70**, 206–14.

FAIRBRIDGE, R. W. [1961], Eustatic changes in sea level; *Physics and Chemistry of the Earth*, **4**, 99–185.

FLOHN, H. [1953], Studen über die atmosphärische Zirkulation in der letzten Eiszeit; *Erdkunde*, **7**, 266–75.

HYLCKAMA, T. E. A. VAN [1956], *The Water Balance of the Earth*, Publications in Climatology, 9, No. 2, Drexel Institute of Technology (Centerton, New Jersey), 117 p.

JELGERSMA, S. [1966], Sea-level changes during the last 10,000 years; In Sawyer, J. S., Editor, *World Climate from 8000 to 0 B.C.*, Royal Meteorological Society (London), pp. 54–69.

L'VOVICH, M. I. [1961], The water balance of the land; *Soviet Geography*, **2**, No. 4, 14–28.

L'VOVICH, M. I. [1962], The water balance and its zonal characteristics; *Soviet Geography*, **3**, No. 10, 37–50.

MALKUS, J. S. [1962], Inter-change of properties between sea and air. Large-scale interactions; in Hill, M. N., Editor, *The Sea*, volume 1 (New York), pp. 88–294.

MILLER, D. H. [1965], The heat and water budget of the earth's surface; *Advances in Geophysics*, **11**, 175–302.

MÖLLER, F. [1951], Vierteljahrskarten des Niederschlags für die ganze Erde; *Petermann's Geographische Mitteilungen*, **95**, 1–7.

PEIXOTO, J. P. and CRISI, A. E. [1965], *Hemispheric Humidity Conditions during the IGY*, Scientific Report No. 6, Meteorology Department, Massachusetts Institute of Technology (Cambridge, Mass.).

RASMUSSON, E. M. [1967], Atmospheric water vapor transport and the water balance of North America; *Monthly Weather Review*, **95**, 403–26.

SCHUMM, S. A. [1965], Quaternary palaeohydrology; In Wright, H. E., Jr., and Frey, D. G., Editors, *The Quaternary of the United States* (Princeton, New Jersey), pp. 783–94.

SELLERS, W. D. [1965], *Physical Climatology* (Chicago), 272 p.

SHUMSKIY, P. A., KRENKE, A. N., and ZOTIKOV, I. A. [1964], Ice and its changes; In Odishaw, H., Editor, *Research in Geophysics*, volume 2 (Cambridge, Massachusetts), pp. 425–60.

SUTCLIFFE, R. C. [1956], Water balance and the general circulation of the atmosphere; *Quarterly Journal of the Royal Meteorological Society*, **82**, 385–95.

TUCKER, G. B. [1961], Precipitation over the North Atlantic Ocean; *Quarterly Journal of the Royal Meteorological Society*, **87**, 147–58.

2.I. The Basin Hydrological Cycle

ROSEMARY J. MORE

Formerly of Department of Civil Engineering, Imperial College, London University

1. Introduction

The river basin, bounded by its drainage divide and subject to surface and subsurface drainage under gravity to the ocean or to interior lakes, forms the logical areal unit for hydrological studies (figs. 2.1.1 (*a* and *b*)). Within this framework one can conveniently, for example, draw up a water balance and assess water resources; estimate the probability of the occurrence of extreme events, such as floods and droughts, particularly as they affect reservoir storage and water use by man; and mobilize hydrological information to enable man to manage his water resources more efficiently by knowing when and in what ways it is to his advantage to intervene locally in the hydrological cycle. It is the purpose of this section to single out the basin as a proper basis for such studies; to identify basin inputs, storages, transfers, and outputs; and to indicate some of the different methods of studying the relationships between these parameters.

The basin cycle can be viewed simply as inputs of precipitation (p) being distributed through a number of storages by a series of transfers, leading to outputs of basin channel runoff (q), evapotranspiration (e), and deep outflow of ground water (b) (fig. 2.1.1(c)). For all practical purposes, the last output is difficult to measure and, except under special geological conditions, is usually assumed to be of such small relative importance as to be commonly ignored in basin input/output studies. Thus the gross operation of the basin hydrological cycle may be simply approximated as:

Precipitation = Basin channel runoff + Evapotranspiration + Changes in storage (1)

or

$$p = q + e + \Delta(I, R, M, L, G, S)$$

The operation of the basin cycle may now be considered in more detail. Precipitation (p) (in the form of rain, sleet, hail, and dew – the delayed hydrological effects of snow will be considered in Chapter 10.1) falls on vegetation, bare rock, debris, and soil surfaces, as well as directly into bodies of standing water and stream channels. Water in transit is stored on the vegetation leaf and stem surfaces as interception storage (I), which either evaporates (e_i) or reaches the ground by stem flow and drip (i). The drainage of water from vegetation, together with direct precipitation on to the ground surface and surface water,

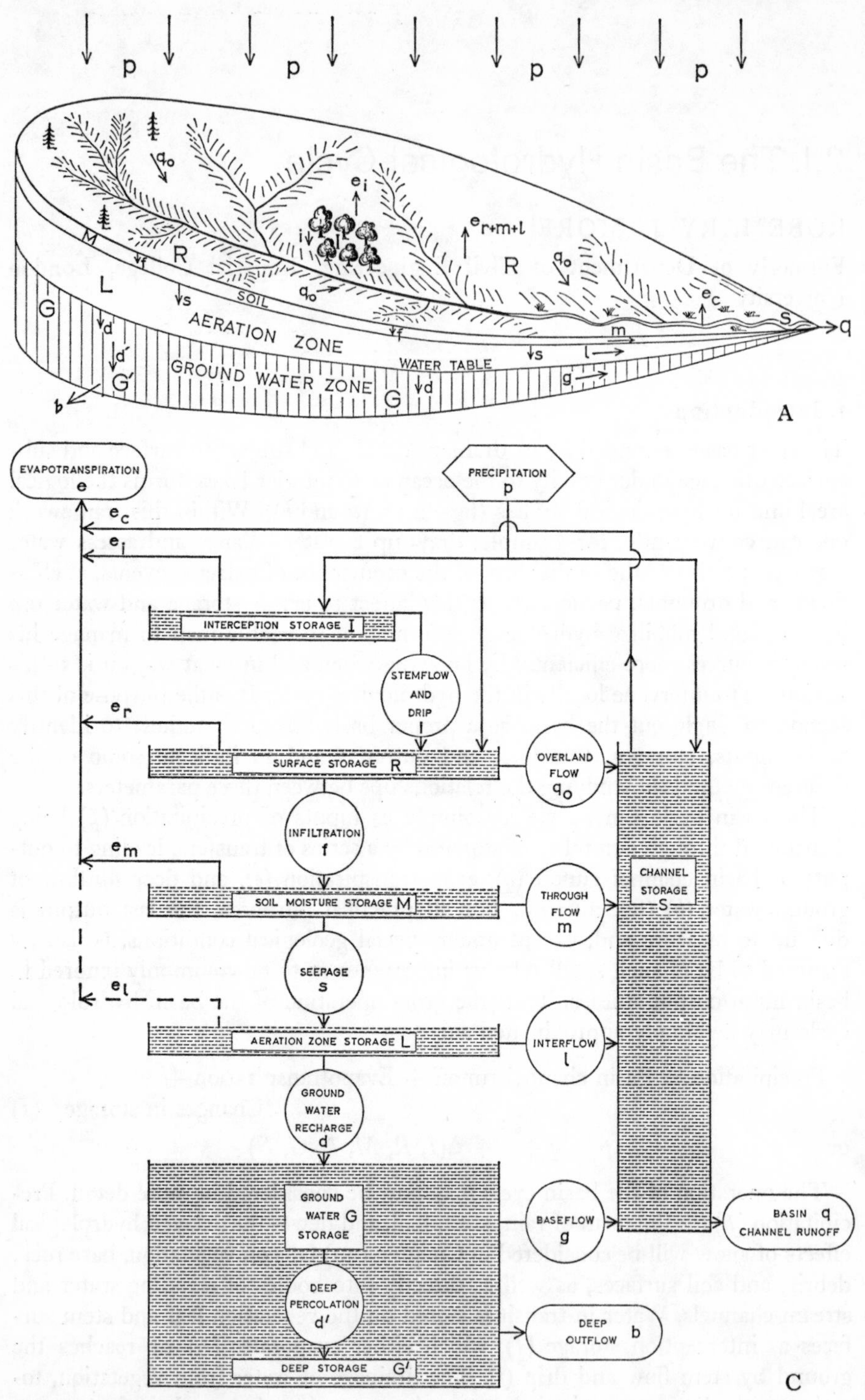

Fig. 2.1.1 The components of the basin hydrological cycle.

A. Block diagram of the basin.
C. Schematic inter-relationships of the basin components.

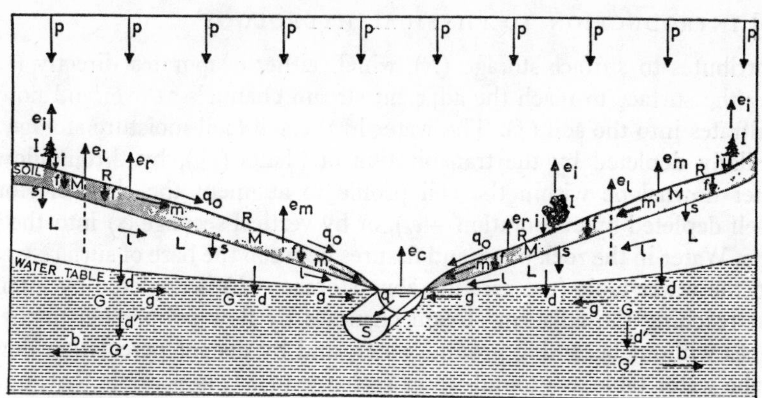

B

TOTAL EVAPOTRANSPIRATION e
RECORDED PRECIPITATION p
CATCHMENT AREA DISTRIBUTION SUBSYSTEM
VEGETATION SUBSYSTEM
INTERCEPTION STORAGE I
STEMFLOW AND DRIP i
SURFACE SUBSYSTEM
SURFACE STORAGE R
OVERLAND FLOW q_o
INFILTRATION f
SOIL SUBSYSTEM
SOIL MOISTURE STORAGE M
THROUGH FLOW m
SEEPAGE s
AERATION ZONE SUBSYSTEM
AERATION ZONE STORAGE L
INTERFLOW l
GROUND WATER RECHARGE d
GROUND WATER SUBSYSTEM
GROUND WATER STORAGE G
BASEFLOW g
DEEP PERCOLATION d′
DEEP STORAGE G′
DEEP OUTFLOW b
CHANNEL SUBSYSTEM
CHANNEL STORAGE S
BASIN CHANNEL RUNOFF (SIMULATED) q_S
MODIFY SUBSYSTEMS
COMPARE
NO
YES
BASIN CHANNEL RUNOFF (RECORDED) q_R
PREDICTED PRECIPITATION p
ADJUSTED MODEL
BASIN CHANNEL RUNOFF (PREDICTED) q_P
e_c e_I e_r e_m e_l

Key
input
output
transfer
storage
subsystem
instruction

D

B. Cross-section of the basin.
D. Flow diagram of the basin components.

contributes to surface storage (R), which either evaporates directly (e_r), flows over the surface to reach the adjacent stream channels as overland flow (q_o), or infiltrates into the soil (f). The water in the soil (soil moisture storage – M) is similarly depleted by the transpiration of plants (e_m), by throughflow (m) of water downslope within the soil profile to augment the channel storage (S) (itself depleted by evaporation —e_c), or by vertical seepage (s) into the aeration zone. Water in the rock pores and fissures between the base of surface weathering and the water table (the aeration zone storage – L) is depleted by interflow (l), that reaching the adjacent stream channels by flow subparallel to the surface slopes without becoming ground water, and by deeper percolation downwards to the water table as ground-water recharge (d). It is possible that, under prolonged dry conditions, appreciable evapotranspiration takes place from the aeration zone (e_l). The deep percolation enters the ground-water storage (G), from which water either flows laterally into the stream channels as baseflow (g), or slowly percolates (d') into deep storage (G'), some of which may be ultimately destined to form deep outflow (b). The latter may discharge at depth into the ocean far from the location of the original precipitation, or may augment the ground-water storage of an adjacent basin. Of course, all the water entering a given basin storage zone is not necessarily released over a short time period and basin water-budget studies must take into account changes in water storage. For example, after a long dry period soil-moisture storage may be so depleted that there may be virtually no seepage or throughflow associated with the first subsequent rainstorm. Thus, through this series of storages and transfers, the incoming precipitation may be related to basin channel runoff, evapotranspiration, and changes in storage, as shown in equation (1).

The components of the basin hydrological cycle and the relationships between them have been studied in five main ways: by natural analogues, hardware models, synthetic systems, partial systems, and the 'black box' approach.

2. Natural analogues and hardware models

A natural analogue may be defined as some analogous natural system believed to be simpler, better known, or in some respects more readily observable than the original. In river-basin hydrology *representative* and *experimental* basins provide such natural analogues, which form the bases for understanding and predicting the behaviour of other, often larger, basins (UNESCO, 1964).

The aim of establishing representative basins is that they should represent the hydrological operation of basins of similar geometry, geology, climate, soils, vegetation, and land use. For example, the Trent River Authority has eighteen such basins, including Bradwell Brook–Peakshole Water in Derbyshire, intended to be representative of Carboniferous Limestone areas, and the River Rea basin in central Birmingham, representative of urban areas. The role of experimental basins, on the other hand, is to provide data for a study of the hydrological response of the basin to artificially-induced physical changes during the period of experimentation. Such changes might include, for example, deforestation or other variations of land use. Of course, the proper interpretation of these data requires that the whole basin output can be rationalized in terms of the input, and this feature of experimental, as well as of representative, basins of 'needing to know the answer before you can do the sum' is one of the most severe restrictions on their use. There are other difficulties in deciding whether the natural analogies are valid ones; both in that the storm precipitation inputs into a small basin used as an analogue have effects different from similar inputs into larger basins (More, 1967, p. 166) and because the large number of basin parameters means that one is never quite sure that there is complete accordance between the natural analogue and the other basins, the behaviour of which one wishes to predict with its aid.

One obvious way of trying to avoid this last difficulty is to attack the basin hydrological cycle by means of *hardware models*, wherein important structural elements of the basin are physically constructed either as scale models (involving such real world materials as sand and water) or analogue models (where there is a radical change in the media used to represent the basin elements, e.g. the representation of the flow of water by the flow of heat or electricity). A laboratory catchment usually partakes of the features of both scale and analogue models in an attempt to predict the runoff output resulting from a given precipitation input, controlled in space and time. Such catchments have been constructed by the U.S. Department of Agriculture (Chery, 1966) and in the Hydraulics Laboratory at Imperial College, London. The advantages of hardware models are that the basin hydrological cycle can be simplified by only considering the controlled variables pertinent to a particular problem and by the ability of the operator to compress natural time sequences into comparatively short experiments. However, many hydrologists (e.g. Amorocho and Hart, 1965) emphasize the difficulties in attempting to simulate natural basin behaviour from a hardware model, pointing to the impossibility of establishing complete dynamic similitude between the model and the prototype, and to the impossibility of operating the many simulated variables over a sufficiently wide range for adequate simulation. To such workers the electronic computer represents the only acceptable hardware model.

3. Synthetic systems

Nash [1967], for example, considers that progress in understanding the basin hydrological cycle will be made most rapidly by the use of overall *synthetic systems*, analysed by computers. In approaching the operation of the basin cycle viewed as a synthetic system, the investigator attempts to describe the operation of its hydrological cycle by a linkage or combination of components, the presence of which is assumed to exist in the system and whose functions are known and predictable (Amorocho and Hart, 1964). The qualitative specification of the basin components and their relationships can be organized into a flow chart, such as is shown in fig. 2.1.1(d), from which computer programmes can be prepared.

The input into the system may be in the form, for example, of hourly rainfall amounts recorded for a number of stations throughout the basin. Each of these rainfall recorders is strategically situated in a segment of the basin so as to

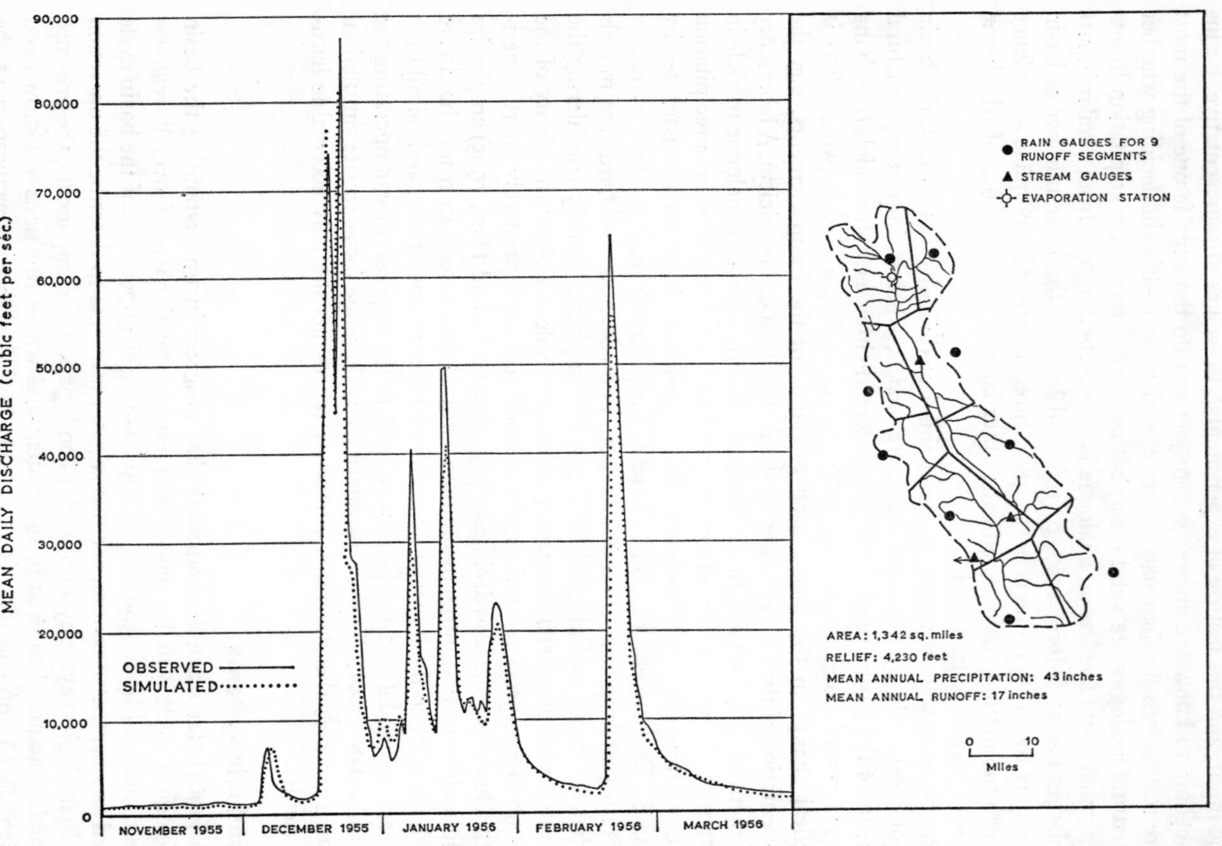

Fig. 2.1.2 Mean daily flow of the Russian River basin at Guerneville, California, November 1955–March 1956, recorded and simulated by the Stanford Watershed Model IV (After Crawford, N. H. and Linsley, R. K., *Dept. of Engineering Stanford University, Technical Rept. 39, 1966*).

represent the input for that segment, and the subsequent basin analysis is largely based on the assumed behaviour of the rainfall input into and through these segments (fig. 2.1.2).

Before the basin system is ready to receive the time sequence of segment inputs from a real basin it must be programmed. This is done by:

1. Inserting estimates of the initial segment storages in each subsystem at the start of the rainfall input sequence.
2. Feeding in an appropriate channel time-delay histogram giving estimated times of surface flow from each of the contributing basin segments along the channel system to the outflow gauging station.
3. Specifying a number of parameters (i.e. mathematical constants) which control and allocate the distribution of precipitation inputs into and between the basin subsystems through time. Some of these parameters are simple invariates (e.g. the measured areas of the basin segments), others can be virtually specified by rule of thumb (e.g. the maximum interception storage for a given vegetation), but many of the parameters are complex time relationships (e.g. the rate of infiltration; the rate of baseflow from ground water). The optimizing of these last parameters is one of the most complex parts of the whole operation.

As the sequences of precipitation inputs are fed into the computerized system the programming described above allows them to filter through the system mathematically, contributing to storage changes, evapotranspiration, and basin channel runoff. (Deep outflow is usually ignored, except in rare geological circumstances where it is significant). The main output from this system is a simulated channel runoff sequence from the basin outlet, and this is compared with the actual recorded channel runoff sequence for the appropriate period. If the two fail to reach some acceptable level of agreement the computer is programmed so that it will automatically modify the subsystem parameters in a predetermined manner, re-run the inputs, and continue this process until the required level of agreement between the simulated and actual basin channel runoff sequences is achieved (fig. 2.1.1(d)). Figure 2.1.2 shows actual and simulated runoffs by the Stanford Watershed Model IV for the Russian River basin above Guerneville, California, for the period November 1955 to March 1956. (For the computer flow diagram and further description see fig. 9.1.5.)

Once the model is working well its potential uses are most important. For example, expected future rainstorm inputs can be used to yield simulated predicted runoff outputs of significance in terms of flood-control planning, and the system parameters can be varied at will to anticipate the hydrological consequences of man-made changes to the basin, such as extensive forest clearance, cultivation, or building construction.

4. Partial systems and the 'black box'

It is clear that the example just given represents research into the basic operation of all the assumed major components of the basin hydrological cycle, in order to

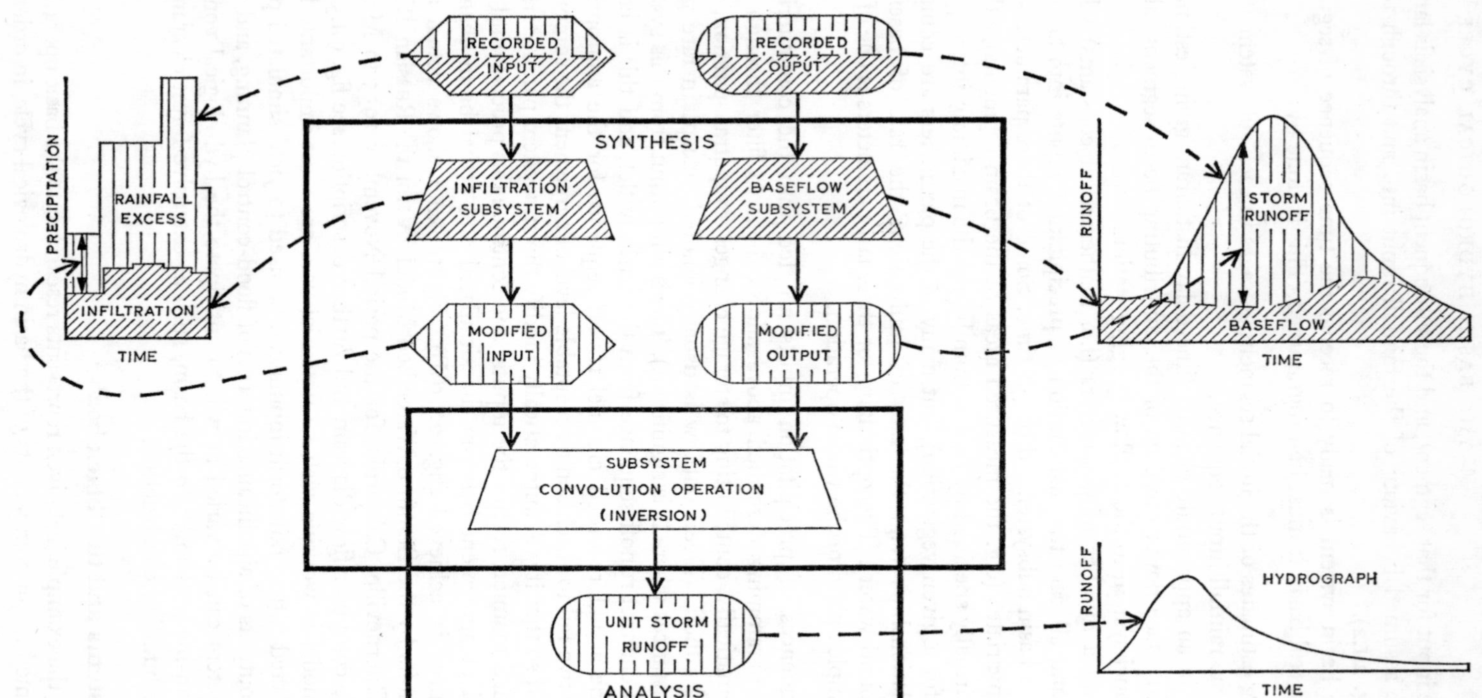

Fig. 2.1.3 Flow chart for partial system synthesis involving the prediction of runoff from precipitation characteristics, with a minimum knowledge of the internal operations of the system components (From Amorocho and Hart, 1964).

gain an understanding of their mechanism and interactions. For some practical purposes, however, it is unnecessary to attempt this degree of analysis, and the *partial-systems* approach depends for its operation on more limited assumptions about the basin cycle. An example of the partial-systems approach, which is also computer based, has been given by Amorocho and Hart [1964] (fig. 2.1.3). It is assumed that a knowledge of only one system parameter is required for the precipitation input (i.e. the proportion of rainfall which infiltrates during the given storm) and one parameter for the output (the proportion of basin discharge contributed by surface runoff, as distinct from baseflow). The operation of the partial system is based on the assumption that rainfall excess over infiltration can be equated with the storm runoff component of the hydrograph, and that the infiltration makes up the baseflow. A given recorded storm input is thus treated in the infiltration subsystem to disentangle the assumed rainfall excess from the infiltration, and the recorded runoff output is similarly treated in the baseflow subsystem to disentangle the assumed surface runoff from the baseflow. The resulting calculated rainfall excess and calculated surface runoff are then compared in the convolution subsystem and the initial assumptions (i.e. rainfall excess/infiltration and surface runoff/baseflow relationships) are modified until a required measure of accordance has been reached. It is then hoped that any storm precipitation pattern through time can be input into the programme to yield very important information relating to the possible river runoff resulting from it.

The last method of approach to understanding the basin cycle is by the use of '*black-box*' techniques. The basin cycle is considered to be a black box, where little or no detailed knowledge is assumed regarding the components or relationships within the cycle and interest is entirely focused on inputs (e.g. rainfall) and outputs (e.g. runoff) and in establishing some direct functional link between them. Because of the complexity of the cycle and the variations of its components from time to time and from area to area it is very difficult to find general mathematical expressions relating rainfall and runoff for a given basin, and little progress has yet been made in this approach.

These five methods of approach to the hydrological cycle are interrelated, and progress in any branch is usually of value in another branch of research, the method chosen being generally dependent on the aims of the investigator.

REFERENCES

AMOROCHO, J. and HART, W. E. [1964], A critique of current methods in hydrologic systems investigation; *Transactions of the American Geophysical Union*, **45**, 307–21.

AMOROCHO, J. and HART, W. E. [1965], The use of laboratory catchments in the study of hydrological systems; *Journal of Hydrology*, **3**, 106–23.

CHERY, D. L. [1966], Design and tests of a physical watershed model; *Journal of Hydrology*, **4**, 224–35.

DAWDY, D. R. and O'DONNELL, T. [1965], Mathematical models of catchment behavior; *Proceedings of the American Society of Civil Engineers, Journal of the Hydraulics Division*, **91**, No. H 74, Part I, 123–37.

MORE, R. J. [1967], Hydrological models and geography; In Chorley, R. J. and Haggett, P., Editors, *Models in Geography* (Methuen and Co., London), pp. 145–85.

NASH, J. E. [1967], The role of parametric hydrology; *Journal of the Institution of Water Engineers*, **21** (5), 435–56.

UNESCO [1964], *Document NS/188, International Hydrological Decade, Inter-Governmental Meeting of Experts, Final Report.*

WOLF, P. O. [1966], Comparison of methods of flood estimation; *The Institution of Civil Engineers, Proceedings of the Symposium on River Flood Hydrology*, 1–23.

2.II. The Drainage Basin as the Fundamental Geomorphic Unit

R. J. CHORLEY
Department of Geography, Cambridge University

1. Morphometric units

The need for the precise description of the geometry of landforms, particularly those of dominantly fluvial erosive origin, has been a recurring theme in geomorphology, and one of the most important aspects of this has been the search for the basic areal unit within which these data could be collected, organized, and analysed. The conceptions of the nature of these units have been very much a product of the broader methodological approaches to geography and earth science in general, and to geomorphology in particular, and can be grouped into three categories. The first important approach (Fenneman, 1914) sprang from the interest of geographers a half century ago in regional delimitation. The physiographic regions so delimited for the United States were based largely upon considerations of structural geology (e.g. the Ridge and Valley province), although certain gross morphometric attributes, notably relief and degree of dissection, were also used. The modern equivalent of this approach is provided by the terrain analogues of the U.S. Corps of Engineers, who used four terrain factors (characteristic slope, characteristic relief, occurrence of steep slopes greater than 26·5°, and the characteristic plan profile involving the 'peakedness', areal extent, elongation, and orientation of topographic highs) to divide up the *gross landscape* of a region into *component landscapes* in a simple taxonomic manner (fig. 2.II.1). In contrast with this basis, the second approach was concerned to identify 'the physiographic atoms out of which the matter of regions is built' (Wooldridge, 1932, p. 33). These 'atoms', however, were defined as the *facets* of 'flats' and 'slopes' forming the intersecting surfaces characteristic of polycyclic landscapes (Wooldridge, 1932, pp. 31–3), and, although this doctrinaire definition has been relaxed to include *segments* of smoothly curved surface (Savigear, 1965) and to allow the grouping of facets into *landscape patterns*, such as a 'mature river valley' (Beckett and Webster, 1962) (fig. 2.II.1), the genetic overtones and subjective character of this morphometric division limits its usefulness (Gregory and Brown, 1966). The third basis for morphometric division results from the obvious unitary features both of geometry and process exhibited by the erosional drainage basin, as recognized long ago by Playfair (Chorley, Dunn and Beckinsale, 1964, pp. 61–3) and by Davis [1899], who wrote:

> Although the river and the hill-side waste sheet do not resemble each other at first sight, they are only the extreme members of a continuous series, and when this generalization is appreciated, one may fairly extend the 'river' all over its basin and up to its very divides. Ordinarily treated, the river is like the veins of a leaf; broadly viewed, it is like the entire leaf.

This topographic, hydraulic, and hydrological unity of the basin provided the basis for the morphometric system of R. E. Horton [1945], as elaborated by Strahler [1964], and it is now employed as a basic erosional landscape element because it is:

1. A limited, convenient, and usually clearly defined and unambiguous topographic unit, available in a nested hierarchy of sizes on the basis of stream ordering.
2. An open physical system in terms of inputs of precipitation and solar radiation, and outputs of discharge, evaporation, and reradiation (Lee, 1964).

2. Linear aspects of the basin

The defining of the perimeter of a drainage basin in the above terms is not difficult, especially as in the majority of instances the ground-water divides are coincident with the topographic ones, but more problems are presented in the definition of the stream-channel network. Definition is especially difficult for the fingertip tributaries in regions of deep soil and plentiful vegetation, whereas in arid shale badlands it is also difficult to distinguish between permanent channels infrequently occupied by runoff and ephemeral rills. The definition of a stream segment, either from the map or in the field, involves five considerations: for a given region or map scale it must have a lower limiting size; it must be connected with the main stream network; it must be 'permanent', as distinct from seasonal; it must form part of a distinctly bifurcating channel pattern; and it must conduct laterally concentrated surface runoff from a well-defined drainage area. A further problem is that the heads of distinct channels are constantly migrating in response to storm excavation or prolonged infill of slope debris (Kirkby and Chorley, 1967).

The linear aspects of stream networks can be analysed from two main viewpoints:

(*a*) the *topological*, which considers the interconnections of the system and yields some scheme of stream ordering; and
(*b*) the *geometrical*, having to do with the lengths, shapes, and orientations of the constituent parts of the network.

The recognition of a hierarchy of stream segments is important because of the different morphometric and hydrologic features associated with each. The most widely used ordering scheme was adapted by Strahler [see, for example, 1964] from Horton [1945], in which fingertip channels are specified as order (U) 1, and where two first-order tributaries join, a channel segment of order 2 is formed,

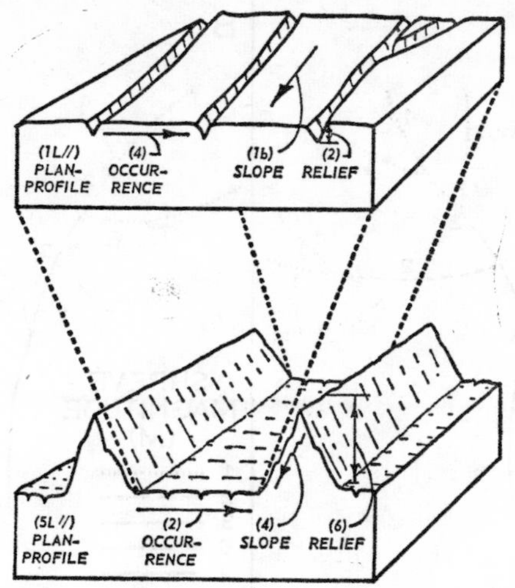

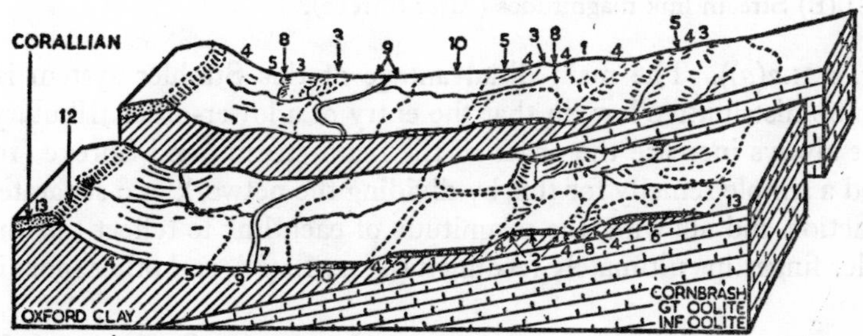

Fig. 2.II.1 Landscape units and geomorphic regions.

Above: Example of a component landscape defined in terms of four terrain factors, and the relation between a component (*top*) and a gross landscape (From Van Lopik and Kolb, 1959).

Below: The pattern of a mature river valley developed by the Upper Thames on the Oxford Clay, illustrating the facets and their relation to each other in the landscape (From Beckett and Webster, 1962).

Facets of river valley and clay

1. High gravel terrace.
2. Spring line.
3. Clay crest.
4. Clay slope.
5. Clay footslope.
6. Unbedded glacial drift.
7. River and banks.
8. Local bottomland.
9. Flood plain alluvium.
10. Old alluvium, not flooded.

Facets of scarplands bounding the river valley pattern

12 (11) Scarp slope.

13. Dipslope.

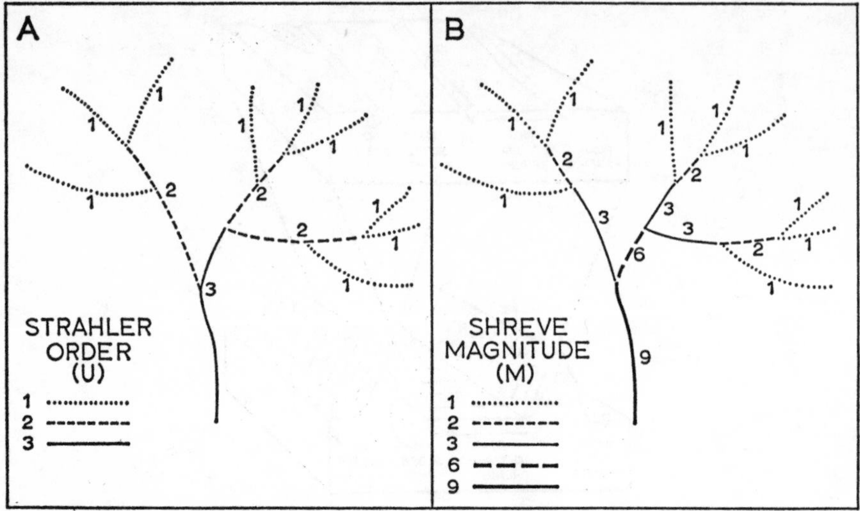

Fig. 2.11.2 Two methods of stream network ordering: (A) Stream segment orders (After Strahler); (B) Stream link magnitudes (After Shreve).

etc. (fig. 2.11.2(*a*)). The main disadvantage of this Strahler system is that it violates the distributive law, in that the entry of a lower-order tributary stream does not always increase the order of the main stream, and Shreve [1966] has proposed a simple remedy for this by dividing the network into separate links at each junction and allowing the magnitude of each link to reflect the number of first-order fingertips ultimately feeding it (fig. 2.11.2(*b*)), and other more involved

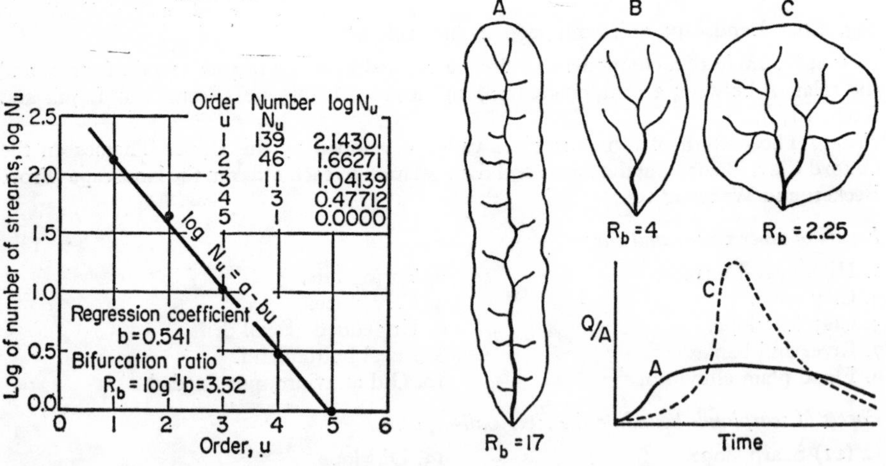

Fig. 2.11.3 The bifurcation ratio (From Strahler, 1964).
Left: Plot of number of stream segments versus order, with a fitted regression.
Right: Hypothetical drainage basins of differing bifurcation ratios, together with their extreme effects on the runoff hydrograph.

schemes have been suggested. However, the simpler unambiguous Strahler system is now firmly established, and this ordering system provides sequences of stream order numbers (N_1, N_2, . . . N_K) which approximate an inverse geometric series for a given basin with the degree of branching, or bifurcation ratio (R_b), given by the ratios N_1/N_2, N_2/N_3, etc., or the antilog of

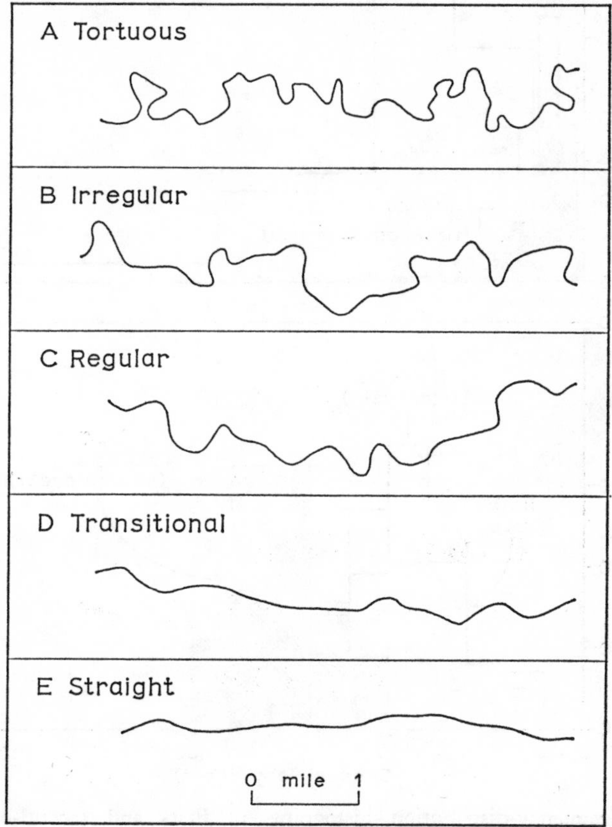

Fig. 2.II.4 Examples of channel patterns (From Schumm, S.A., 1963, *Bulletin of the Geological Society of America*). (A) White River near Whitney, Neb. ($P = 2{\cdot}1$); (B) Solomon River near Niles, Kan. ($P = 1{\cdot}7$); (C) South Loup River near St. Michael, Neb. ($P = 1{\cdot}5$); (D) North Fork Republican River near Benkleman, Neb. ($P = 1{\cdot}2$); (E) Niobrara River near Hay Springs, Neb. ($P = 1{\cdot}0$).

the regression coefficient (b) (fig. 2.II.3). The bifurcation ratio, for a given density of drainage lines, is very much controlled by basin shape and shows very little variation (ranging between 3 and 5) in homogeneous bedrock from one area to another. Where structural effects cause basin elongation, however, this value may increase appreciably. Besides influencing the landscape morphometry, the bifurcation ratio is an important control over the 'peakedness' of the runoff hydrograph (fig. 2.II.3) (see Chapter 9.I).

The ratio between the measured length of a stream channel and that of the

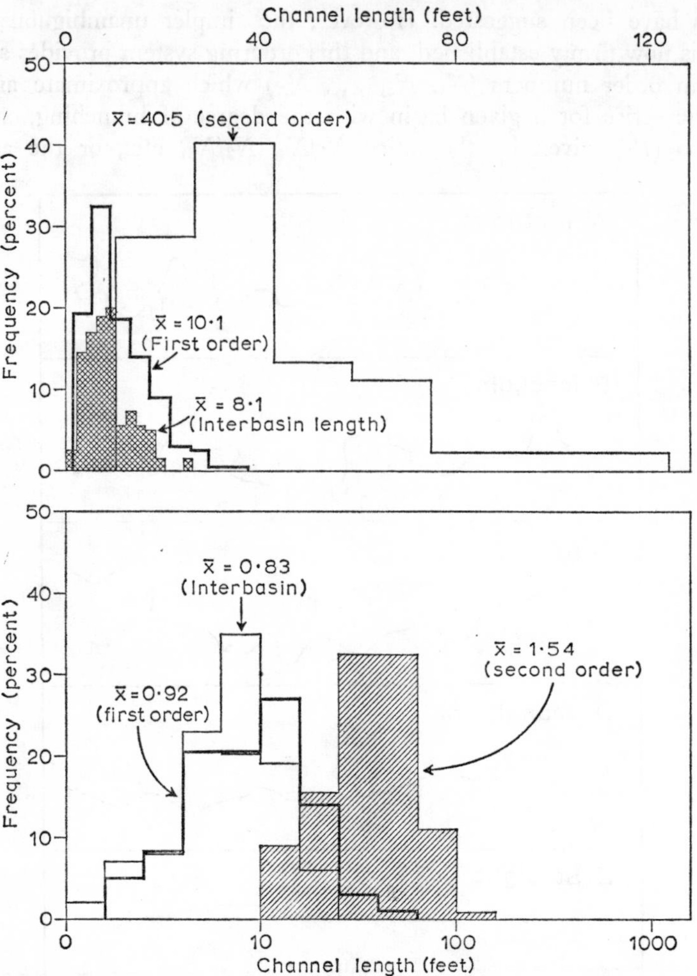

Fig. 2.II.5 Frequency-distribution histograms of first- and second-order channel lengths and maximum interbasin lengths for the Perth Amboy Badlands, New Jersey (From Schumm, 1956).

Above: Actual stream lengths.
Below: Logarithms of stream lengths.

thalweg of its valley is a measure of its sinuosity (fig. 2.II.4). Distributions of lengths of streams of each order in a drainage basin are characteristically right-skewed (log-normal) (fig. 2.II.5), and the plot of mean stream lengths of each order ($\bar{L}_1$, $\bar{L}_2$, $\bar{L}_3$, . . . $\bar{L}_K$) in a basin produces an approximation to a direct geometric series (fig. 2.II.6), where the antilog of the regression coefficient is the length ration (R_l).

Obviously the absolute length of the channel system exercises a strong control over the basin lag time (the time difference between rainfall and the resulting

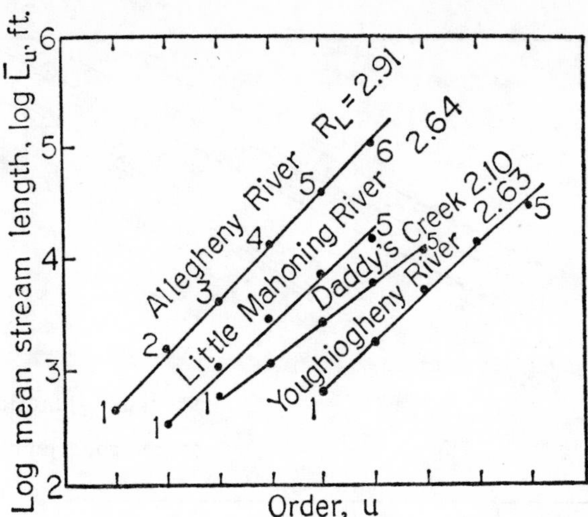

Fig. 2.11.6 Regression of logarithm of mean stream segment length versus order for four drainage basins in the Appalachian Plateau Province (After Morisawa; From Strahler, 1964).

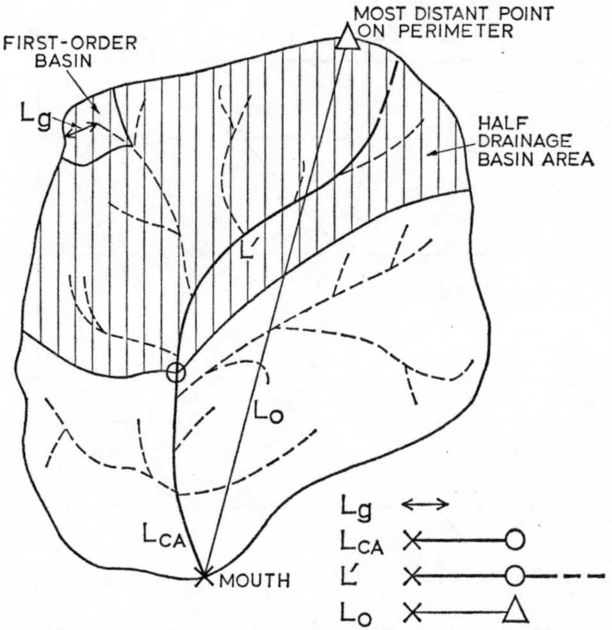

Fig. 2.11.7 Some common drainage basin length parameters.

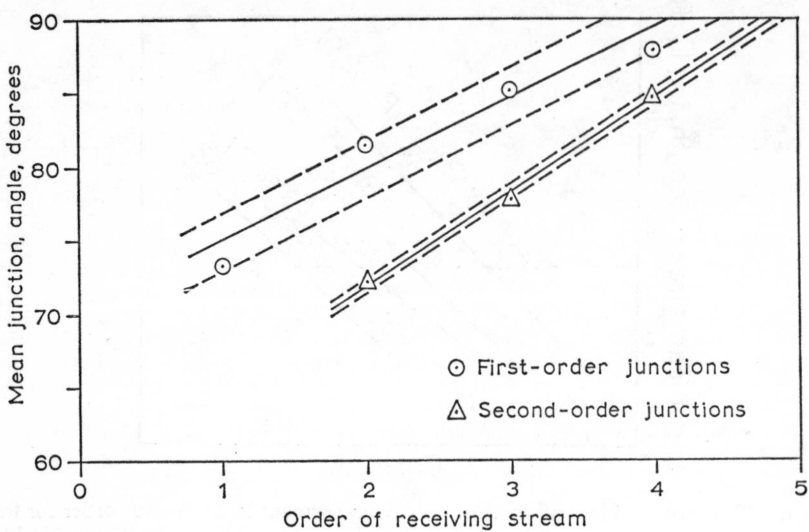

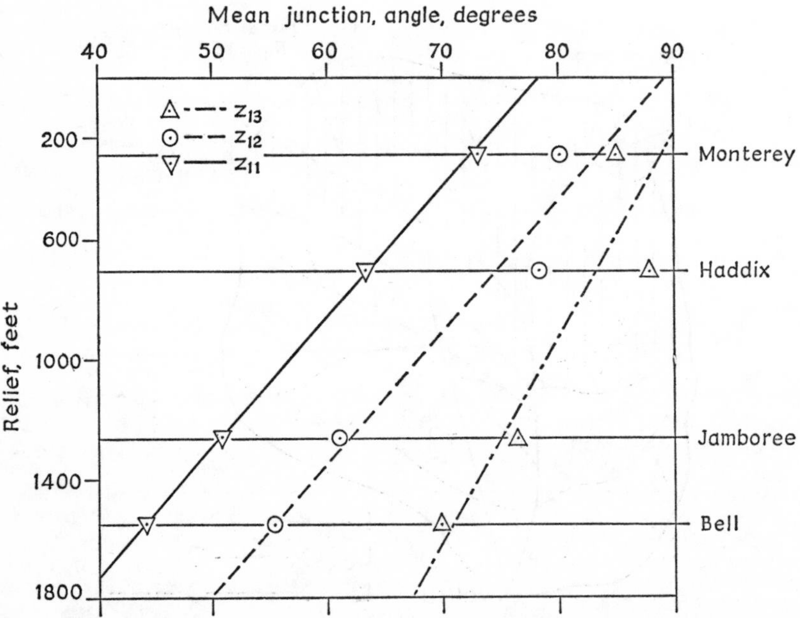

Fig. 2.11.8 Relationships of mean stream junction angles (From Lubowe, J. K., 1964, *American Journal of Science*).

Above: Plot of mean junction angle of first- and second-order with receiving streams of similar or higher order, in part of the Lexington Plain, Kentucky.

Below: Plot of mean junction angles of first-order streams with first-, second-, and third-order receiving streams in four areas of the east, central and western United States.

stream runoff: see Chapter 9.1), as do the following length parameters (fig. 2.II.7):

1. The length of overland flow (L_g), which is the mean horizontal length of the flow path from the divide to the stream in a first-order basin. This parameter is a measure of stream spacing, or degree of dissection, and is approximately one-half the reciprocal of the drainage density $\left(L_g \approx \frac{1}{2D}\right)$. As the mean velocity of unconcentrated overland flow is less than $\frac{1}{5}$ that of concentrated channel flow (30,000 cm/hr, versus 160,000 cm/hr), L_g is also an important control over lag time. A less-meaningful measure of stream spacing is the mean stream interval (MI), calculated from sampling stream intersections with a grid.
2. The length of the main stream of the basin (L') (usually designated by following up from the mouth those streams which make the least angle with the next lower segment). This is sometimes continued to the basin margin (then called the 'mesh length').
3. The distance to the 'centre of gravity' of the drainage basin (L_{ca}). This is usually measured up the main stream to a point where one half of the drainage basin area lies headward of it, but for most basins $L_{ca} = 0{\cdot}5L$ is a good approximation. It is sometimes measured to the centre of gravity of the basin area.
4. The length of the longest basin diameter (L_o), measured from the basin mouth to the most distant point on the perimeter. This is useful in the calculation of basin shape.

The entrance angle (Z_c) between a tributary developed in a valley-side slope (of $\theta°$) and joining a larger stream of lower slope ($\gamma°$) would be approximately expressed by:

$$\cos Z_c = \frac{\tan \gamma}{\tan \theta}$$

As the slopes are degraded through time, one might expect θ to approach γ, and consequently Z_c to decrease. It is characteristic that mean values of junction angle increase as the order of the receiving stream increases (i.e as the difference between γ and θ increases), and it is inversely related to relief (for given orders of junction), probably because high relief imparts especially high gradients to the receiving streams (fig. 2.II.8).

3. Areal aspects of the basin

Basin area (conventionally referred to a horizontal datum plane) is hydrologically important because it directly affects the size of the storm hydrograph and the magnitudes of peak and mean runoff (fig. 2.II.9) (See Chapter 9.1). It is interesting that the maximum flood discharge per unit area is inversely related to size, because the most intense storms are usually of the smallest size (fig. 2.II.9) (More, 1967, p. 166). In a given large drainage basin developed in a homogeneous region

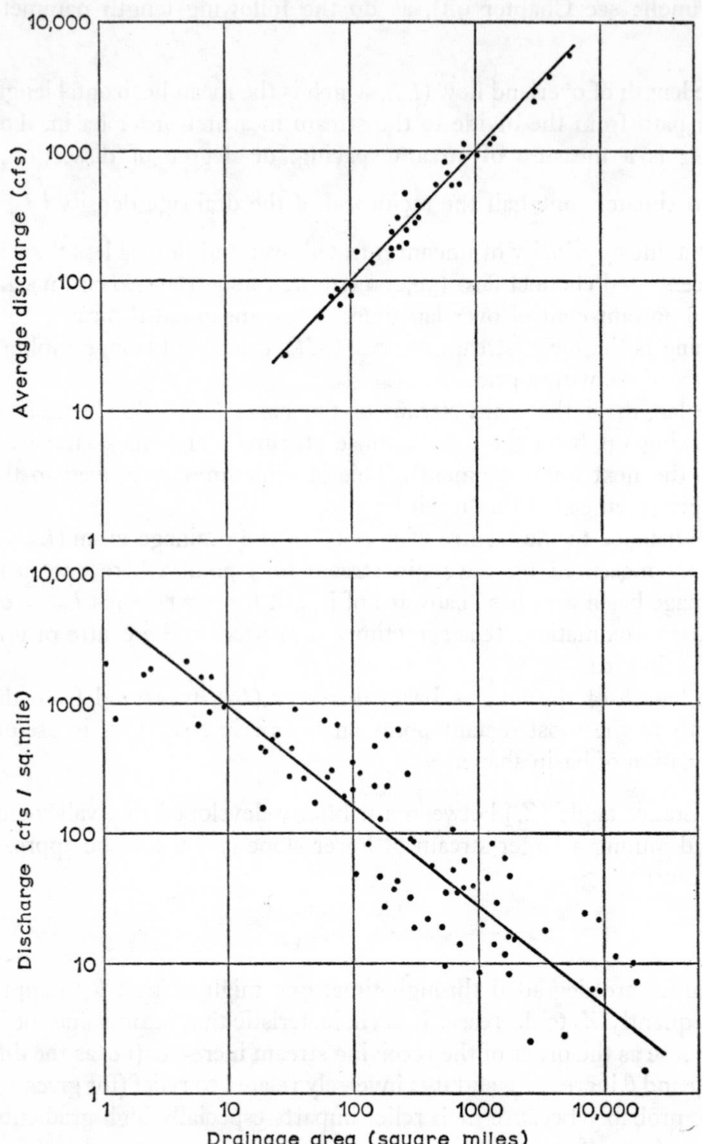

Fig. 2.II.9 Relations between basin area and stream discharge.

Above: Mean discharge (cfs) versus drainage area for all gauging stations on the Potomac River (After Hack: From Strahler, 1964).

Below: Maximum flood discharge (cfs) per square mile versus drainage area for basins in Colorado (From Follansbee, R. and Sawyer, L. R., 1948, *U.S. Geological Survey Water Supply Paper* 997).

basin areas of given order show a logarithmic-normal distribution (fig. 2.II.10), the means of which approximate a direct geometric series (fig. 2.II.11). By relating characteristic discharge to drainage area (the relationship $Q_{2\cdot33} = 12A^{0\cdot79}$ was obtained for basins in central New Mexico, where $Q_{2\cdot33}$ is the flood discharge equalled or exceeded on average once every 2·33 years, or the mean annual

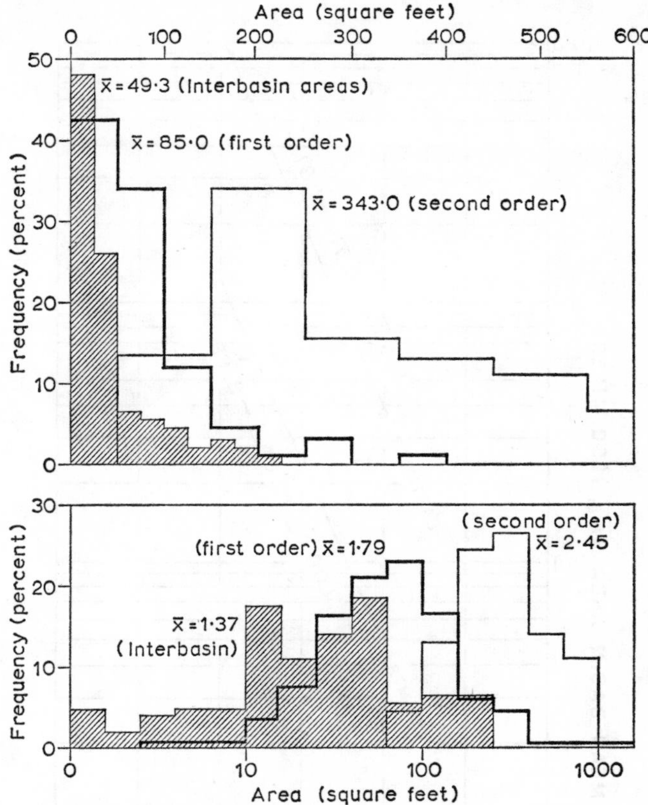

Fig. 2.II.10 Frequency-distribution histograms of first- and second-order basin areas and interbasin areas for the Perth Amboy Badlands, New Jersey (From Schumm, 1956).
Above: Actual basin areas.
Below: Logarithms of basin areas.

flood: see Chapter 7.II), and then area to order, it is possible to show an exponential relationship between order and discharge (fig. 2.II.12).

The relationship of stream length to basin area is important because plots of basin area draining into various locations along the main stream (i.e. area-distance curves) give an idea of the pattern of runoff (fig. 2.II.13), and also because the relationship of total stream lengths of all orders to basin area is one of the most sensitive and variable morphometric parameters, and one which controls the texture of landscape dissection and the spacing of streams. Thus drainage density (*D*), for example, is defined as the total stream length per unit area of basin (e.g. in miles per square mile). Drainage density exhibits a very wide range of values in nature and is commonly believed to reflect the operation of the complex factors controlling surface runoff. Common values of *D* are, for example, 3–4 in the sandstones of Exmoor and the Appalachian plateaus,

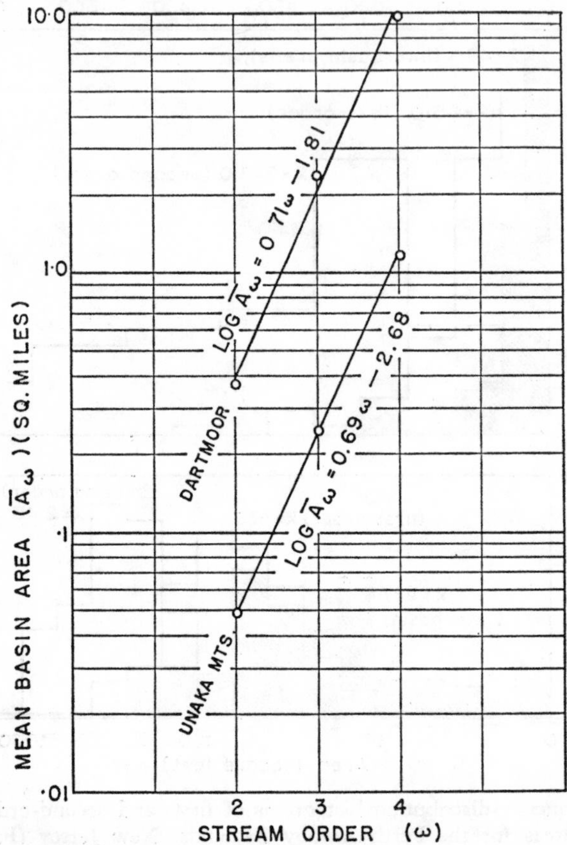

Fig. 2.II.11 Regressions of mean area versus order for drainage basins in the Unaka Mountains, Tenn. and N. Car., and Dartmoor, England (From Chorley, R. J. and Morgan, M. A., 1962, *Bulletin of the Geological Society of America*).

20–30 for the scrub-covered Coast Ranges of California, 200–400 for the shales of the Dakota badlands, and up to 1,300 for unvegetated clays (fig. 2.II.14). An allied measure is the constant of channel maintenance which is the area (in square feet) necessary to maintain 1 ft of drainage channel. Drainage density affects the runoff pattern, in that a high drainage density removes surface runoff rapidly, decreasing the lag time and increasing the peak of the hydrograph (see Chapter 9.1). Of course, basin shape (itself largely controlled by geological structure) is an important control over the geometry of the stream network (fig. 2.II.15). Examples of simple shape measures are:

1. The circularity ratio – (see Chapter 9.1).
2. The elongation ratio – the diameter of a circle having the same area as the basin, as a ratio of the maximum basin length (L_0).
3. The measure $(L' \cdot L_{ca})^{0\cdot3}$, found by experience to be a good predictor of basin lag.

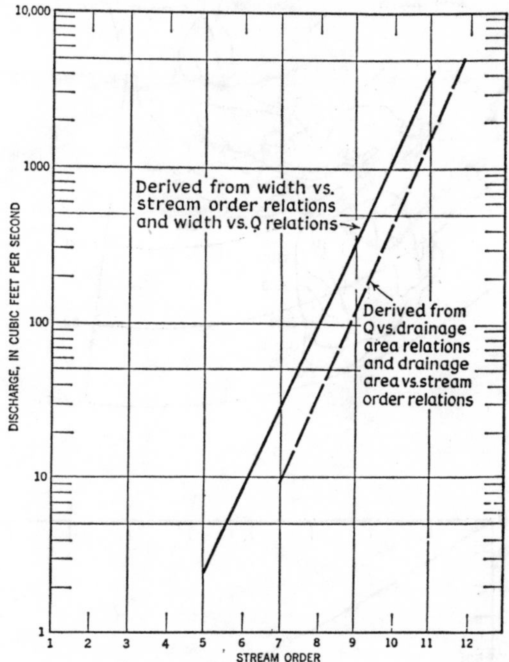

Fig. 2.II.12 Relation of discharge to stream order for ephemeral streams in New Mexico derived by two separate types of analysis (From Leopold, L. B. and Miller, J. P., 1956, *U.S. Geological Survey Professional Paper* 282–A).

It is interesting that, unless pronounced structural control is present, drainage basins differ relatively little in shape, although basins tend to become more elongate with strong relief and steep slopes.

4. Relief aspects of the basin

Longitudinal profiles of stream channels are characteristically concave-up, and it has been suggested that this is due to the increase of discharge downstream not being balanced by any commensurate increase in frictional losses – the increase in the wetted perimeter being more than compensated for by the increase in cross-sectional area and by the decrease of bed material grain size (due to sorting and abrasion during transportation). Many attempts have been made to fit

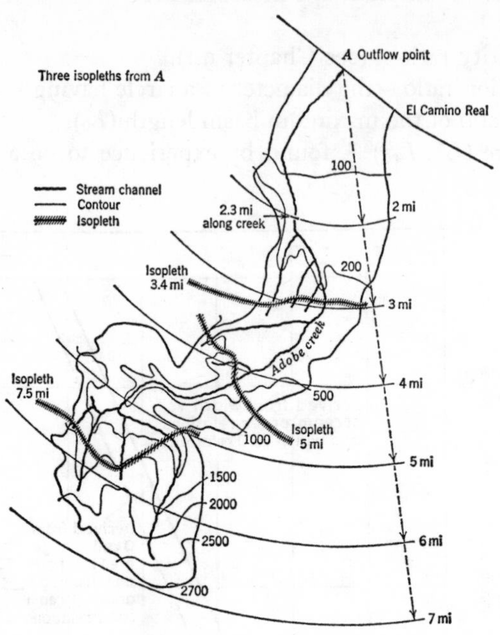

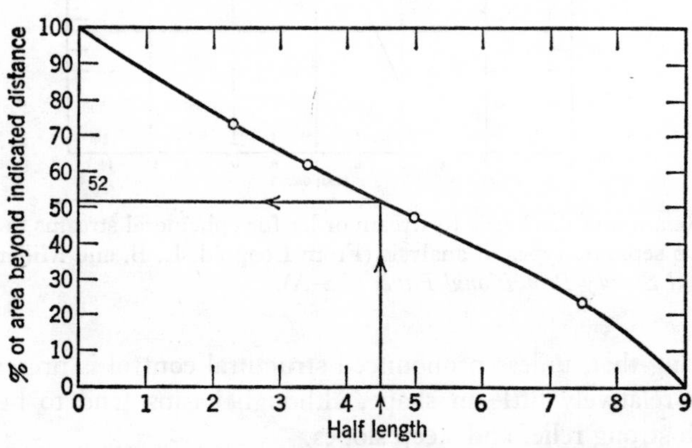

Fig. 2.11.13 Area as a function of channel distance from the basin mouth for Adobe Creek, near Palo Alto, California (Area 11 square miles; drainage density 2·18 miles/square mile) (From De Wiest, 1965).

Above: Adobe Creek, showing channel distance isopleths.

Below: Distribution of area as a function of distance from the basin mouth.

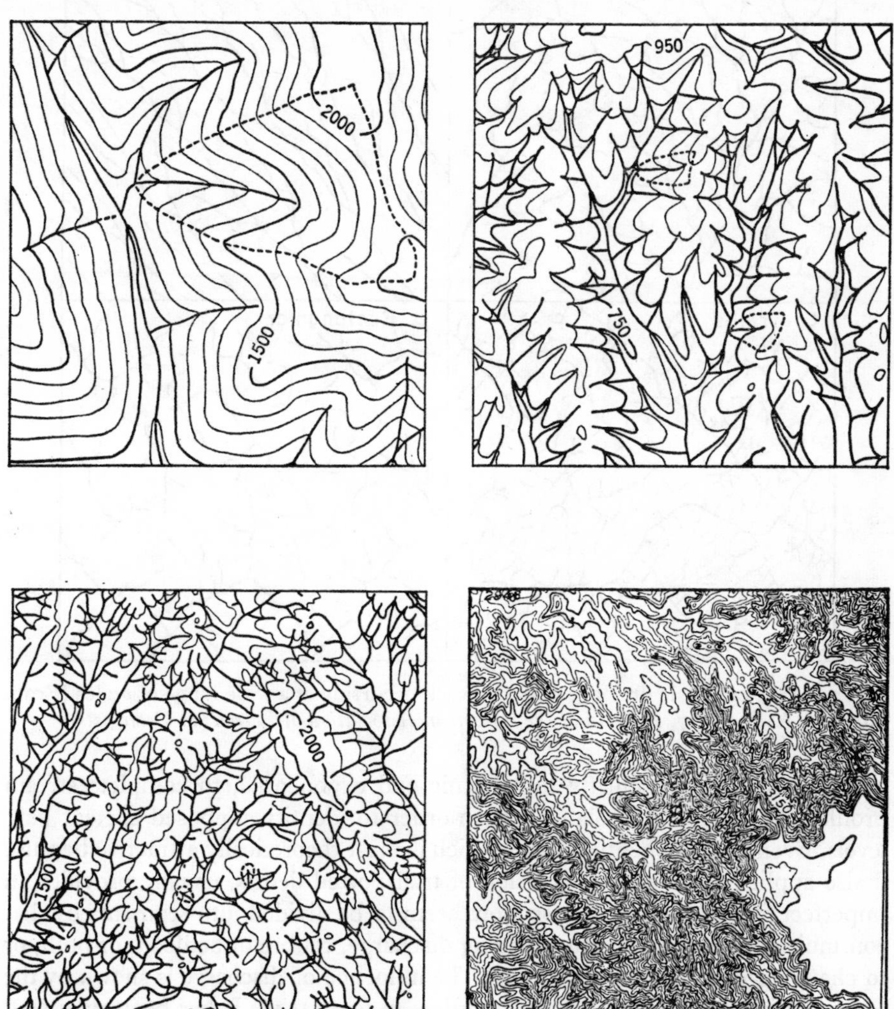

Fig. 2.II.14 Four areas, each of 1 square mile, illustrating the natural range of drainage density (From Strahler, 1964).

Top left: Low drainage density: Driftwood Quad., Penn.
Top right: Medium drainage density: Nashville Quad., Ind.
Bottom left: High drainage density: Little Tujungo Quad., Cal.
Bottom right: Very high drainage density: Cuny Table West Quad., S. Dak.

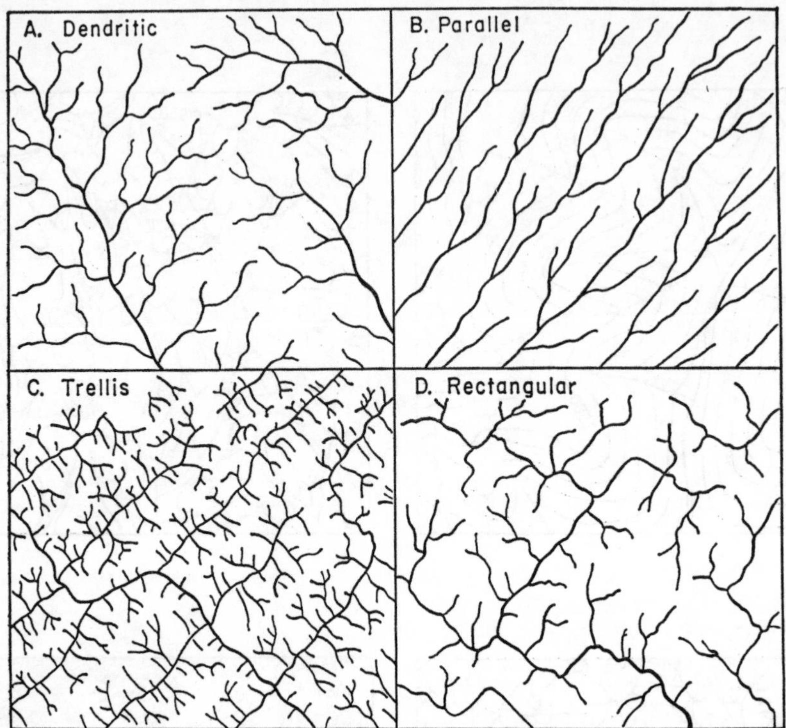

Fig. 2.II.15 Four basic drainage patterns, each occurring at a wide range of scales (From Howard, A. D., 1967, *Bulletin of the American Association of Petroleum Geologists*).

mathematical curves, chiefly logarithmic and exponential, to complete stream profiles, largely to extrapolate them in order to identify supposed ancient base levels. Even the exponential curve, which is most theoretically attractive because it also applies to the rate of attrition of transported debris, usually provides an imperfect fit, and long stream sections seem to be divided into segments by discontinuities which are due to changing discharge, calibre of bedload, or both, or to changing channel characteristics. The influence of discharge is shown by the segmentation on the basis of stream orders in fig. 2.II.16. It has been suggested that in a given basin mean channel gradient bears an inverse geometric relationship to stream order (fig. 2.II.17), although the imperfection of this relationship has led to other more complex ones being suggested. For practical purposes other stream-gradient measures are employed, including:

1. Some measure of the average slope of the main channel in a basin. This is commonly the arithmetic mean slope ($\bar{S}$) of the whole channel, or the slope of the 'equivalent' stream (i.e. one having the same length and flood peak travel time) (S_{st}).
2. The simple gross slope of the channel, obtained by dividing the elevation difference between head and mouth by the length of the main stream.

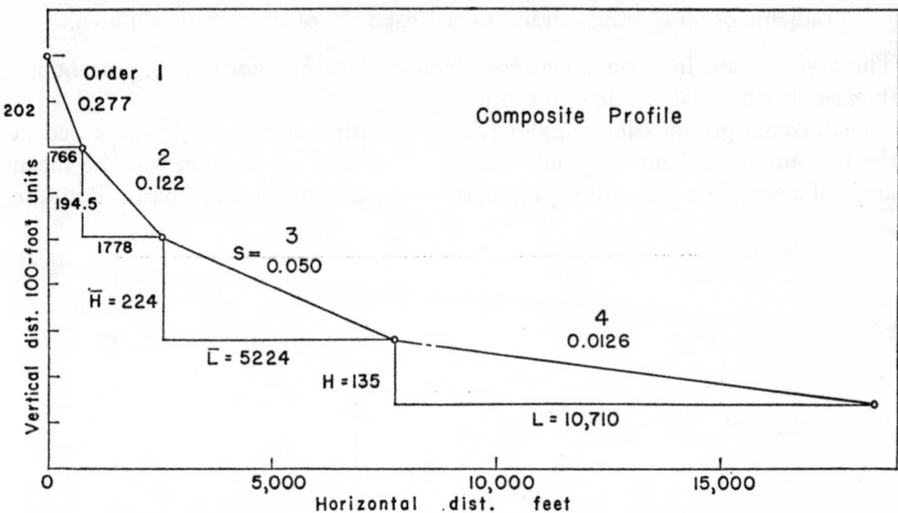

Fig. 2.11.16 Plot of the longitudinal profile of Salt Run, Penn., showing the difference in mean slope of each of the four segments of differing order (From Broscoe, 1959, *Office of Naval Research Technical Report 18, Project NR 389–042, Contract N6 ONR 271–30*).

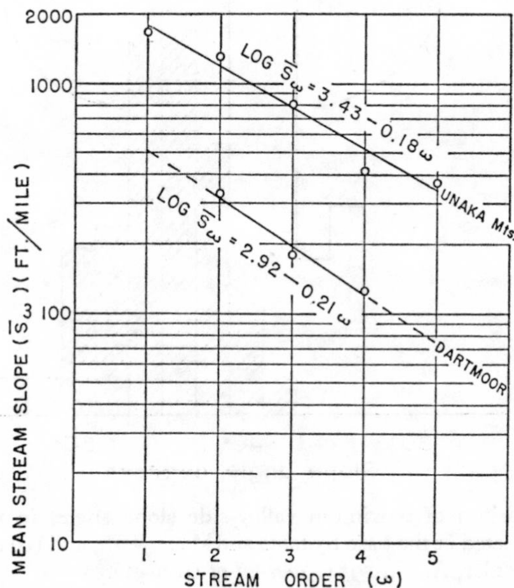

Fig. 2.11.17 Regressions of mean channel slope versus order for streams in the Unaka Mountains, Tenn. and N. Car., and Dartmoor, England (From Chorley, R. J. and Morgan, M. A., 1962, *Bulletin of the Geological Society of America*).

3. The mean slope of the whole channel system, computed by averaging the gradients of all channels draining at least 10% of the total basin area.

These slopes are important because channel slope exercises an important influence over the magnitude of the runoff peak (see Chapter 9.1).

Orthogonal ground-slope angles (S_g) in a basin are commonly measured at the maximum gradient of given valley-side profiles, or expressed as the mean angle of a complete valley-side profile, or sampled over the whole basin. Both the

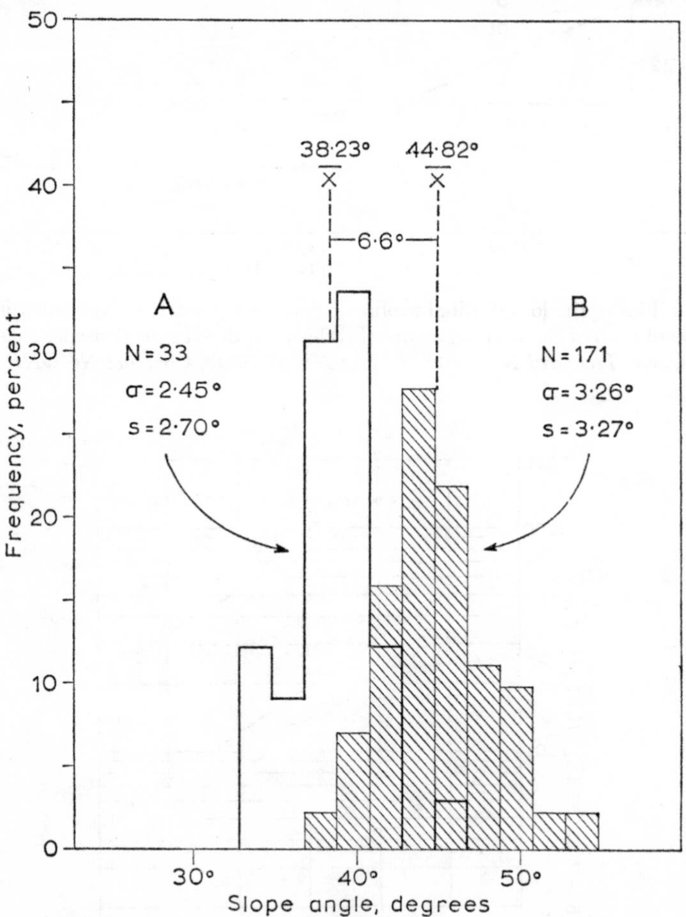

Fig. 2.11.18 Comparison of maximum valley-side slope angles in the Verdugo Hills, California, (A) protected at the base by talus and slope wash, and (B) actively corraded at the base (From Strahler, A. N., 1954, *Journal of Geology*).

maximum and mean valley-side slope angles within a basin are commonly normally distributed (fig. 2.11.18), and, although they are not individually related to the gradient of the basal stream, a mean relationship seems to exist when different regions are compared. The character of the distribution of slope angles sampled over the whole basin depends on the height distribution within it; for a 'just mature' basin with limited flat summit or floodplain areas the distribution is normal, but other basins give skewed distributions, the direction of skew depending on whether the small angles are concentrated on the summits ('youth') or on the floodplains ('late maturity'). Again, simple measures of average ground

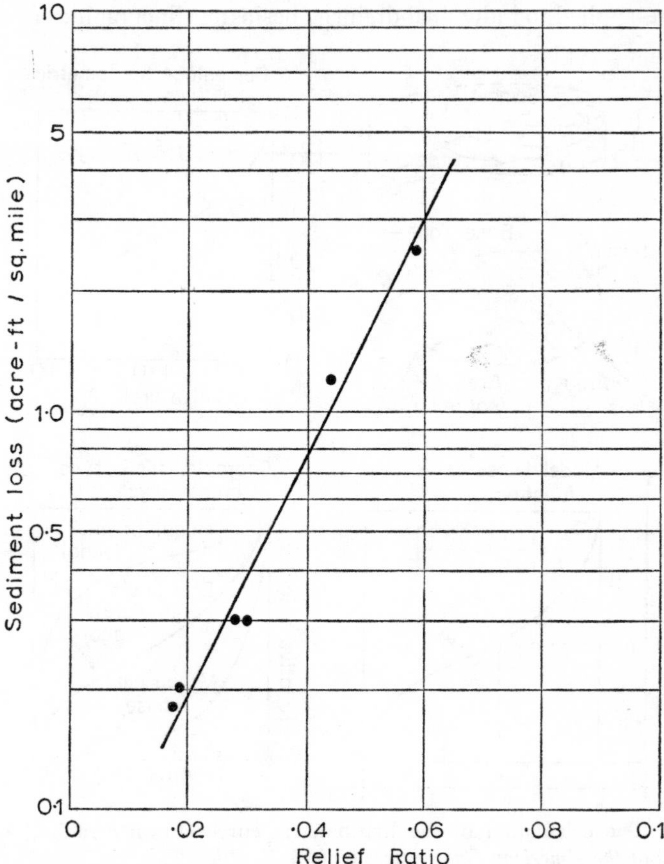

Fig. 2.II.19 Relation of sediment loss to relief ratio for six small drainage basins in the Colorado Plateaus (From Schumm, 1956).

slope within a basin are employed in hydrological analysis, such as the mean basin slope $\left(\frac{\text{Total length of contours} \times \text{Contour interval}}{\text{Basin area}}\right)$, and the relief ratio ($R_h$), which is the ratio of the maximum basin relief to the horizontal distance along the longest basin dimension parallel to the main drainage line. Even such crude measures as these can be used to rationalize basin dynamics, it being found that mean basin slope influences the form of the hydrograph and that the relief ratio exercises an important control over rates of sediment loss from some basins (fig. 2.II.19).

The distributional characteristics of land elevations have long been of geomorphic interest, because concentrations of area with elevation (i.e. surfaces of low slope) were believed to be indicative of ancient base levels. Plots of mean land slope versus elevation (clinographic curves) and of amount of surface area versus elevation (hypsometric curves) were prepared for large upland areas to assist in the evaluation of their possible polycyclic histories. Both techniques have been re-applied to individual drainage basins; a clinographic curve giving a

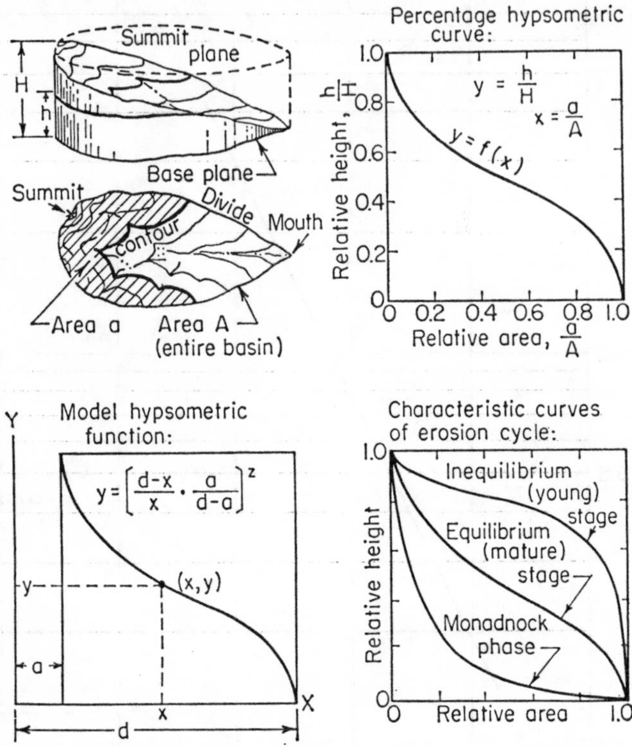

Fig. 2.II.20 The calculation of the hypsometric curve (From Strahler, A.N., 1957, *Transactions of the American Geophysical Union*).

more accurate estimate of land slope than average figures when evaluating the form of the hydrograph, and the dimensionless hypsometric curve representing in some instances the relative stage of basin degradation through time with reference to an assumed original uneroded block (fig. 2.II.20). An especially important hydrological property of the basin related to the distribution of elevations is the amount of floodplain storage available, the effect of which is to make the rising limb of the hydrograph less steep, increase the lag time, and make the peak lower and less pronounced. A knowledge of the distribution of elevations also enables better estimates of rainfall, snowfall, and evaporation in the basin to be made.

In the past, morphometric analysis from maps has been a rather tedious and time-consuming task, but recently techniques have been devised to give it greater facility. The most important of these is the digitizer of the 'pencil-follower' type, which both records the rectangular coordinates of points on a map and also gives a continuous read-out of point coordinates sufficiently closely spaced (maximum recording rate is 20 points per second) to define lines – i.e. contours, stream channels, basin perimeters, etc. The card or tape output can be edited, elaborated, and then fed directly into a computer, together with programmes which will automatically calculate areas, shapes, drainage densities, mean aximuths, maximum slope angles, and the like. Such programming is in its infancy, but already it promises to release the masses of data locked up in topographic maps and will obviously allow much more extensive sampling and generalization of morphometric properties. Before too long these methods will be applied directly to the output from aircraft and satellite scanning equipment, obviating the necessity for the actual compilation of many maps.

5. Some considerations of scale

Considerations of changing scale, both linear and temporal, introduce a certain element of sophistication into studies concerned with morphometric relations.

Similarities such as appear to exist generally between the bifurcation, length, and area ratios of basins of differing size developed on bedrock lacking pronounced structural control have prompted speculation that erosional drainage basins in differing hydrological environments may show a close approximation to geometrical similarity when mean values are considered. If complete geometrical similarity existed one would expect to find all length measurements between corresponding mean points to bear a fixed linear scale ratio, and all corresponding angles to be equal. Although this does not seem to be the case for all morphometric properties (i.e. there appears to be a changing relationship between main-stream length and basin area as basins increase in size within a region, which seems due to both an increasing elongation of larger basins and to the increasing sinuosity of the stream), the similar geometrical relationships between some linear properties suggest a more general significance to the 'laws of morphometry'.

Although it has not been the purpose of this contribution to discuss the relationships between morphometry and erosional processes, it has doubtless become apparent that two quite distinct approaches to this are possible, dependent upon the time-scale with which one is concerned. In the long term it is clear that the hydrological events which compose 'process' must be instrumental in determining the morphometry of the landscape, but, when one is concerned with explaining in the short term the factors which control the character of such individual processes, morphometric features (such as gradient) are commonly invoked. Whether morphometric parameters are viewed as mathematically dependent or independent variables is very much a matter of the time scale employed. Figure 2.II.21 gives two plots involving drainage density, one showing it to be strongly controlled by the precipitation effectiveness and the other

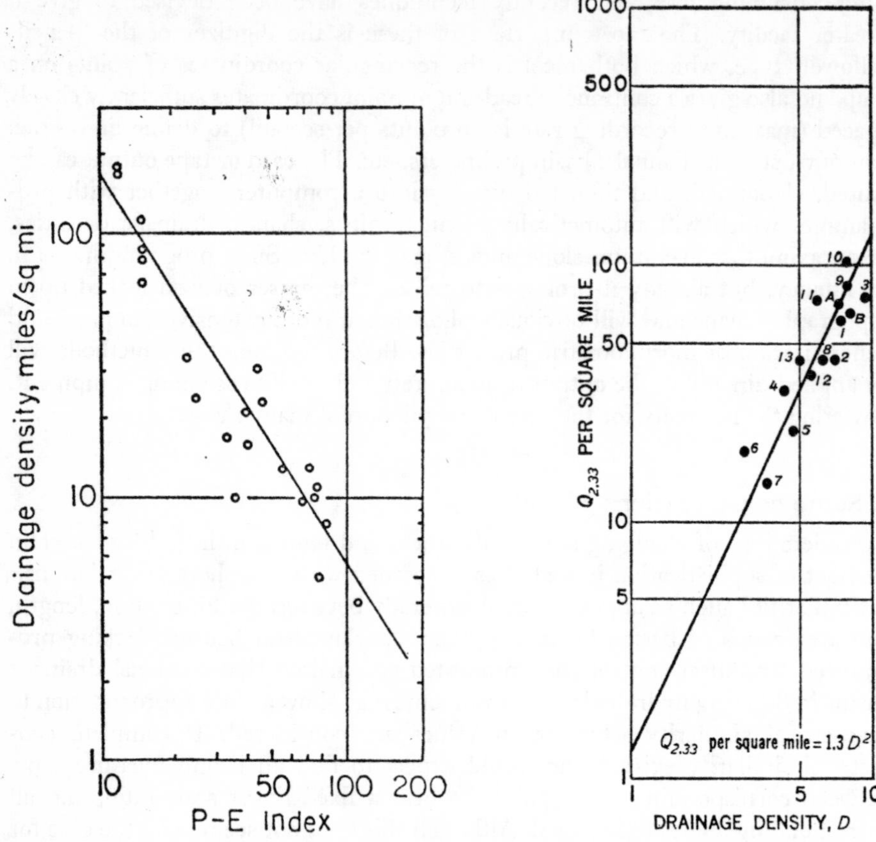

Fig. 2.II.21 The control of drainage density exercised by Thornthwaite's precipitation effectiveness (P–E) index (*Left*) (From Melton, M. A., 1957, *Office of Naval Research Technical Report 16, Project NR 389–042, Contract N6 ONR 271–30*); and (*Right*) the control exercised by drainage density over the mean annual flood ($Q_{2.33}$) for 13 basins in the central and eastern United States (From Carlston, C. A., 1963, *U.S. Geological Survey Professional Paper 422–C*).

indicating how drainage density differences within a region control the peak mean annual flood. A number of the following chapters examine the effects of hydrological processes on aspects of basin morphometry, and Chapter 9.I uses basin morphometry, among other factors, to analyse the form of the flood hydrograph.

REFERENCES

BECKETT, P. H. T. and WEBSTER, R. [1962], The storage and collection of information on terrain (An interim report); *Military Engineering Experimental Establishment, Christchurch, Hampshire*, 39 pp. (Mimeo).

CHORLEY, R. J., DUNN, A. J., and BECKINSALE, R. P. [1964], *The History of the Study of Landforms*, Volume I (Methuen, London), 678 pp.

CLARKE, J. I. [1966], Morphometry from maps; In Dury, G. H., Editor, *Essays in Geomorphology* (Heinemann, London), pp. 235–74.

DAVIS, W. M. [1899], The geographical cycle; *Geographical Journal*, **14**, 481–504.

DE WIEST, R. J. M. [1965], *Geohydrology* (Wiley, New York), 366 pp.

FENNEMAN, N. M. [1914], Physiographic boundaries within the United States; *Annals of the Association of American Geographers*, **4**, 84–134.

GOLDING, B. L. and LOW, D. E. [1960], Physical characteristics of drainage basins; *Proceedings of the American Society of Civil Engineers, Journal of the Hydraulics Division*, **86**, No. HY 3, 1–11.

GRAY, D. M. [1961], Interrelationships of watershed characteristics; *Journal of Geophysical Research*, **66**, 1215–23.

GREGORY, K. J. and BROWN, E. H. [1966], Data processing and the study of land form; *Zeitschrift für Geomorphologie*, Band 10, 237–63.

HORTON, R. E. [1945], Erosional development of streams and their drainage basins: Hydrophysical approach to quantitative morphology; *Bulletin of the Geological Society of America*, **56**, 275–370.

KIRKBY, M. J. and CHORLEY, R. J. [1967], Throughflow, overland flow and erosion; *Bulletin of the International Association of Scientific Hydrology*, Year 12(3), 5–21.

LANGBEIN, W. B. *et al.* [1947], Topographic characteristics of drainage basins; *U.S. Geological Survey Water Supply Paper* 968-C, 125–157.

LEE, R. [1964], Potential insolation as a topoclimatic characteristic of drainage basins; *Bulletin of the International Association of Scientific Hydrology*, Year 9, 27–41.

LEOPOLD, L. B., WOLMAN, M. G., and MILLER, J. P. [1964], *Fluvial Processes in Geomorphology* (Freeman, San Francisco), pp. 131–50.

MORE, R. J. [1967], Hydrological models and geography; In Chorley, R. J. and Haggett, P., Editors, *Models in Geography* (Methuen, London), pp. 145–85.

SAVIGEAR, R. A. G. [1965], A technique for morphological mapping; *Annals of the Association of American Geographers*, **55**, 514–38.

SCHUMM, S. A. [1956], The evolution of drainage systems and slopes in badlands at Perth Amboy, New Jersey; *Bulletin of the Geological Society of America*, **67**, 597–646.

SHREVE, R. L. [1966], Statistical law of stream numbers; *Journal of Geology*, **74**, 17–37.

STRAHLER, A. N. [1964], Quantitative geomorphology of drainage basins and channel networks; In Chow, V. T., Editor, *Handbook of Applied Hydrology* (McGraw-Hill, New York), Section 4–11.

VAN LOPIK, J. R. and KOLB, C. R. [1959], A technique for preparing desert terrain analogs; *U.S. Army Engineer Waterways Experiment Station, Vicksburg, Mississippi, Technical Report* 3–506, 70 pp.

WISLER, C. O. and BRATER, E. F. [1959], *Hydrology*; 2nd edn. (Wiley, New York), 408 pp.

WOOLDRIDGE, S. W. [1932], The cycle of erosion and the representation of relief; *Scottish Geographical Magazine*, **48**, 30–6.

3.I(i). Precipitation

R. G. BARRY

Institute of Arctic and Alpine Research, University of Colorado

1. General categories

Inhabitants of middle latitudes are familiar with four major precipitation categories – drizzle, rain, snow, and hail. For scientific purposes a more detailed classification is necessary (Table 3.1(i).1).

TABLE 3.1(i).1 The major categories of precipitation

Type	*Characteristics*	*Typical amount*
Dew	Deposited on surface, particularly a vegetation canopy (frozen form—hoar frost)	0·1–1·0 mm/night
Fog-drip	Deposited on vegetation and other obstacles from fog (frozen form—rime)	Up to 4 mm/hr
Drizzle	Droplets <0·5 mm in diameter (freezing drizzle when surface temperature below 0° C)	0·2–0·5 mm/hr
Rain	Drops >0·5 mm diameter, typically 1–2 mm diameter	Light <2 mm/hr Heavy >7 mm/hr
Sleet (Great Britain)	Partly melted snow or a rain and snow mixture	
Snowflakes	Aggregations of ice crystals up to several cm across	
Snow grains (granular snow)	Very small, flat opaque grains of ice; the solid equivalent of drizzle	
Snow pellets (graupel, or soft hail)	Opaque pellets of ice 2–5 mm diameter, falling in showers	
Ice pellets (small hail)	Clear ice encasing a snowflake or snow pellet	
Ice pellets (sleet in the U.S.)	Frozen rain or drizzle drops	
Hail	Roughly spherical lumps of ice, 5–50 mm or more diameter, showing a layered structure of opaque and clear ice in cross-section	

2. The precipitation mechanism

Condensation occurs when air is cooled to its dew-point temperature and suitable hygroscopic nuclei are present to initiate droplet formation. Dew-point is the temperature at which saturation occurs if air is cooled at constant pressure. For present purposes it is convenient to differentiate between condensation in the atmosphere and that very close to the ground.

When air rises it is cooled by the adiabatic expansion due to lower pressures, and uplift beyond the level at which condensation occurs leads to the formation of clouds. Precipitation rarely begins until at least 30 minutes after cloud has been observed to form overhead, and many clouds are observed to dissipate without precipitating. This suggests that the growth of cloud droplets, 1–100 microns (1 micron = 10^{-4} cm) in diameter, into rain drops with a diameter of 1 mm or more (1,000 microns) is by no means automatic. In shower clouds with tops not reaching above the freezing level, and more generally in clouds in the tropics, the main mechanism of droplet growth is coalescence. Where ice crystals and supercooled droplets, i.e. liquid at below freezing temperatures, are present in a cloud, the crystals grow at the expense of the droplets because the saturation vapour pressure is lower over ice than over water. Eventually the ice crystals fall from the cloud, and if they pass into air warmer than about 2–3° C they usually melt into rain drops. The microphysics of clouds is treated in most meteorological textbooks; the work of Mason [1962] is specifically concerned with this subject.

Important contributions to the total precipitation are sometimes made by dew and fog. Dew forms on the ground and vegetation surfaces as a result of deposition from the atmosphere when the air is cooled to its dew-point temperature by contact with the radiatively cooling surface. However, in very still air most of the dew originates from vapour derived by evaporation from the soil. The rate of deposition is limited by the rate of removal of latent heat of condensation from the surface. Average dew deposition measured on lysimeters at Coshocton, Ohio, during the snow-free period 1944–55 was between 0·38 and 0·75 mm (0·015 and 0·030 in.) per day.

If the wind stirs the air sufficiently fog droplets are held in suspension. These may accumulate on vegetation and other surfaces as 'fog drip' or, if the drops are supercooled, in frozen form as rime. A fog gauge of wire gauze on Table Mountain, Cape Town, collected 1·7 times more moisture than fell in an ordinary rain gauge during a twelve-month period. Vegetation may be a less efficient collector, but it seems probable that such additions to the moisture budget are important in coastal and montane forests.

3. Genetic types

It is usual to recognize three main types of precipitation, according to the mode of uplift of the air.

A. 'Convective type' precipitation

This occurs in the form of showers or heavy downpours with cumulus and cumulonimbus clouds. Rainfall rates may be of the order of 25 mm (1 in.) per hour. Three subcategories, distinguished according to the spatial organization of the precipitation, are as follows.

1. Summer heating over land causes scattered thundery showers of rain or occasionally hail.
2. Cold, moist air moving over a warmer sea or land surface frequently gives rain (or snow) showers. The convective cells tend to travel with the wind, producing a streaky distribution of precipitation parallel to the wind direction (Bergeron, 1960). The cells may otherwise be organized into bands, some hundreds of kilometres long, perpendicular to the airflow, particularly in association with an advancing wedge of cold air (a cold front or a polar trough).
3. In tropical cyclones cumulonimbus cells become organized about the vortex in spiralling bands of cloud mass. The rainfall can be very heavy and prolonged, affecting large areas. This is often classed as a separate category or as a special case of 'cyclonic type' precipitation. However, a key feature of tropical storms is their mode of energy supply. Energy is generated by the release of latent heat of condensation in cumulonimbus towers. In other words, the small-scale convection maintains the large-scale circulation.

B. 'Cyclonic type' precipitation

In this case the horizontal convergence of airstreams within a low-pressure area brings about widespread upward air motion. In the forward sector of a mid-latitude depression warm air overlies colder air (a warm front) and there is usually deep multi-layered cloud of the nimbostratus type. This gives fairly continuous light to moderate precipitation over very extensive areas. The precipitation may last 6–12 hr or more, according to the width of the rain belt and the speed of the depression. In the rear sector, where cold air tends to undercut warmer air (a cold front, often with associated squall lines preceding the air-mass boundary) heavy showers, quite frequently accompanied by thunder, are usual.

The inadequacy of the simple three-fold classification of precipitation types is shown by the fact that, in the equatorial low-pressure zone, airstream convergence in the easterlies gives rise to westward-moving bands of convective precipitation from cumuliform clouds. This 'cyclonic type' of precipitation could equally well be grouped under A. 2.

C. Orographic precipitation

In the strict sense this term implies that precipitation occurs over high ground when none is falling on the surrounding plains. More frequently, it is a *component* of the total precipitation resulting from the effect of orography on the basic convective and cyclonic mechanisms. The effect is dependent on the size of the barrier and its alignment with respect to the wind. Over narrow uplands the horizontal scale may be insufficient for maximum cloud build-up, and precipitation may be carried over the crest-line by the wind, causing a lee-side maximum.

In middle and higher latitudes it seems that where onshore westerlies are forced to rise sharply over coastal mountains precipitation may increase with height up to 2,000 m (6,500 ft) or more. Walker [1961] considers that maximum precipitation occurs at the level of the cloud base and estimates that in the western Cordillera of British Columbia the maximum precipitation zone occurs as follows (in thousands of feet):

<table>
<tr><th></th><th colspan="2">Coast</th><th colspan="2">Interior</th></tr>
<tr><th></th><th>South</th><th>North</th><th>South</th><th>North</th></tr>
<tr><td>Summer</td><td>6</td><td>4</td><td>7</td><td>6</td></tr>
<tr><td>Winter</td><td colspan="2">4</td><td colspan="2">5</td></tr>
</table>

Farther inland the maximum may occur well below the summit levels. For example, in Norway it is located about half the horizontal distance along the windward slope due to the considerable width of the mountain belt. In Java there is a marked decrease above about 2,000 m, and in Hawaii the maximum of more than 800 cm (320 in.) occurs on the eastern slopes of the mountains at only 1,000 m. Yet on some of the Hawaiian Islands peaks rising to 2,000 m receive their maximum on the summit. The reason for these variations appears to de-

TABLE 3.1(i).2 The occurrence of precipitation types in England and Wales, 1956–60 (after E. M. Shaw and R. P. Mathews)

Station	*Warm front*	*Warm sector*	*Cold front*	*Occlusion*	*Polar low*	mP	cP	*Arctic*	*Thunderstorms*
Cwm Dyli, Snowdonia (324 ft)	18	30	13	10	5	22	0·1	0·8	0·8
Squires Gate Blackpool (33 ft)	23	16	14	15	7	22	0·2	0·7	3
Rotherham, Yorkshire (70 ft)	26	9	11	20	14	15	1·5	1·1	3
Cranwell, Lincolnshire (208 ft)	27	10·5	14	19	9	11	2·0	1·9	5

mP = maritime polar air
cP = continental polar air

pend not only on the vertical distribution of water content, and consequently on the type of cloud system, but also on the precise effects of the mountains on the airflow. The influence of a given mountain range is, of course, markedly affected by the movement of weather systems from different directions. The simple climatological concepts of rainfall increase with height and lee-side rain shadow need to be replaced by more realistic models for a variety of synoptic situations in each mountain area.

Convergence and uplift occur over coastal areas when air moving inland is slowed down by friction. This special type of orographic effect is evident in the patterns of average seasonal precipitation over south-east Sweden, for example, and also on leeward shores of the Great Lakes and Hudson Bay in early winter.

Table 3.1(i).2 provides a more detailed breakdown of annual rainfall at stations in England and Wales during 1956–60. The orographic component is particularly evident for warm-sector precipitation at Cwm Dyli. Air mass showers (mP) diminish in importance eastward across the country.

4. Precipitation characteristics

Basic information about daily precipitation amounts is supplied by rain-gauge and climatological stations. From these data are compiled statistics of average monthly and average annual precipitation, annual variability, and the number of rain-days (⩾0·01 in. or >0·2 mm in Britain). Invaluable as such records are, it is essential in hydrological studies to know more about the characteristics of individual rainstorms. Three important parameters of storm rainfall are intensity, frequency, and areal extent.

A. Intensity

Rainfall intensity (= amount/duration) is of vital interest to hydrologists concerned with flood prevention and conservationists dealing with soil erosion.

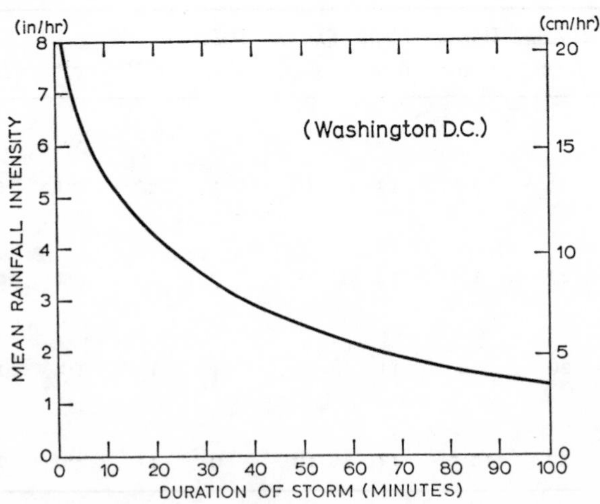

Fig. 3.1(i).1 Generalized relationship between precipitation intensity and duration for Washington, D.C. (After Yarnell, 1935).

Intensity has to be determined from chart records ('hyetograms') of rate-of-rainfall recorders. The results can be presented in the form of an intensity-duration graph as illustrated in fig. 3.1(i).1 for Washington, D.C. Analysis of record precipitation rates from different parts of the world shows that the expected *Global extreme* intensity (in./hr) $\approx \frac{14{\cdot}3}{\sqrt{\text{Duration}}}$. However, this particular assessment overlooked an occurrence of 73·6 in. (187 cm) in 24 hr on the island of Réunion, off Madagascar during March 1952 (Paulhus, 1965). Further discussion is given in Chapter 3.1(i).7.c.

B. Frequency

In many design studies, especially for systems of flood control, it is essential to know the *average* time-period within which a rainfall of specified amount or

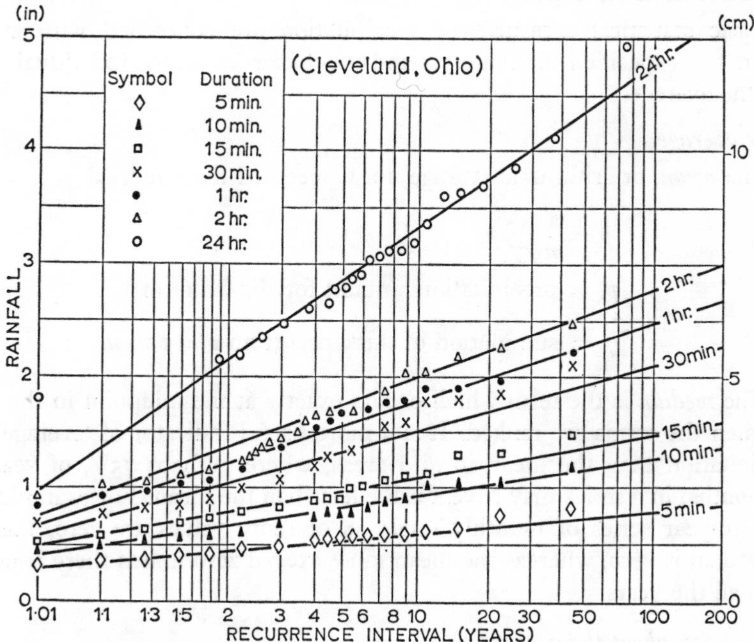

Fig. 3.1(i).2 The return period for annual rainfall totals at Cleveland, Ohio, 1902–47 (After Linsley and Franzini, 1955).

intensity can be expected to occur once. This is known as the 'return period' or 'recurrence interval'. The method of calculation is dealt with in Chapter 11.1.2. Figure 3.1(i).2 illustrates a graph of this type for Cleveland, Ohio.[1]

C. Areal Extent

Storm totals obviously depend on the type and scale of system – local thunderstorm, tropical disturbance or extra-tropical depression – and its rate of move-

[1] A valuable analysis of recent British data is given in *British Rainfall*, 1961 (H.M.S.O. 1967).

ment. In general, the effect of thunderstorm downpours is limited to areas on the scale of individual catchments, whereas steady depression rainfall may affect extensive drainage basins causing large-scale flooding if there is a slow-moving depression. For the United States, maximum 24-hr totals over different areas have been estimated by Gilman [1964] as follows:

ml^2	in.	cm
10 (25·9 km^2)	38·7	98·3
10^2	35·2	89·4
10^3	30·2	76·7
10^4	12·1	30·7
10^5	4·3	10·9

5. Precipitation statistics

The basic statistical measures of precipitation are concerned with average amounts for a specified time interval and the dispersion of the individual values about the average.

A. The average

1. The *mean*, or arithmetic average, for a specified time interval

$$\bar{p} = \frac{1}{n} \sum_{i=1}^{n} p_i$$

where p_i = precipitation amount for the ith term;

$\sum_{i=1}^{n}$ = summation of the terms from $i = 1$ to n.

2. The *median* is the term which occurs exactly at the midpoint in the series when the terms are ranked. It is a more useful indicator of 'average' precipitation than the mean in arid areas, where perhaps 75% of years, or months, in a series may have a value less than the mean. For example, if in a 35-year series of monthly totals 18 or more values are zero, then the median is zero, whereas the mean must exceed zero unless there is no rain in all the years.

B. Dispersion about the average

1. The simplest indicator of dispersion in a series is the *range* between the extremes. In the British Isles the extreme range of annual precipitation at any locality is about 40–180% of the mean, whereas in the arid, south-west United States it is approximately 25–270%. The range increases enormously when shorter time intervals are considered. This measure is not very satisfactory, because it fails to indicate the frequency of a deviation of specified magnitude.
2. Two measures of dispersion are associated with the mean. The simplest to calculate is the *mean deviation*:

$$\text{M.D.} = \sum_{i=1}^{n} |p_i - \bar{p}|$$

where $|p|$ denotes the absolute value of p, without regard to sign. This statistic of dispersion has been widely used in rainfall studies, but it lacks the versatility of the *standard deviation*, σ, in further statistical application, especially the assessment of probabilities.

$$\sigma = \sqrt{\frac{\sum_{i=1}^{n} (p_i - \bar{p})^2}{n}} \quad \text{or} \quad \sqrt{\left\{\frac{\sum_{i=1}^{n} p_i^2}{n} - (\bar{p})^2\right\}}$$

3. In connection with the median it is usual to specify the upper and lower *quartile* values, i.e. at the 75 and 25% positions in the ranked series. The quartiles delimit the central 50% of the frequency distribution. The uppermost and lowermost *deciles* (90 and 10%, respectively) in the series may also be of interest.

C. Relative variability

In order to compare the deviations for places with different average values it is necessary to express them as a percentage of the mean. The simple measure of relative variability (R.V.) is

$$\text{R.V.} (\%) = \frac{\text{M.D.}}{\bar{p}} \times 100$$

This index shows a marked tendency to increase sharply with low annual precipitation totals.

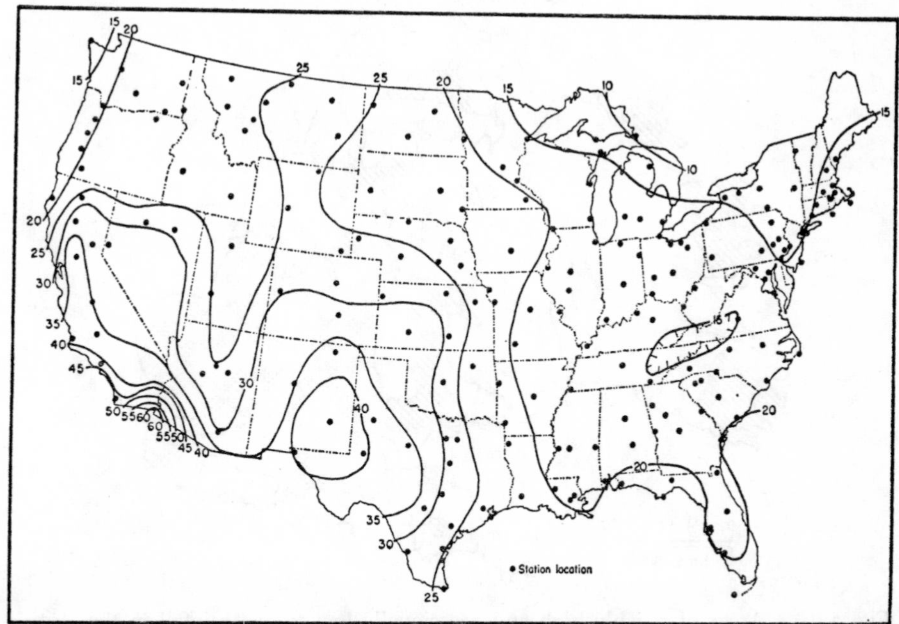

Fig. 3.1(i).3 The coefficient of variation of annual rainfall (%) over the United States (From Hershfield, 1962).

A preferable measure is the *coefficient of variation* (C.V.)

$$\text{C.V.}\ (\%) = \frac{\sigma}{\bar{p}} \times 100$$

In the United States the C.V. for annual precipitation ranges from about 15% in the north-east and 20% in Florida to more than 35–40% in the arid south-west (fig. 3.1(i).3). Both R.V. and C.V. take no account of the sequence of the data, which may be of considerable importance. Wallén uses the following expression for inter-annual variation (I.A.V.):

$$\text{I.A.V.}\ (\%) = 100\ \frac{\sum_{i=1}^{n} |p_{i-1} - p_i|}{n-1}$$

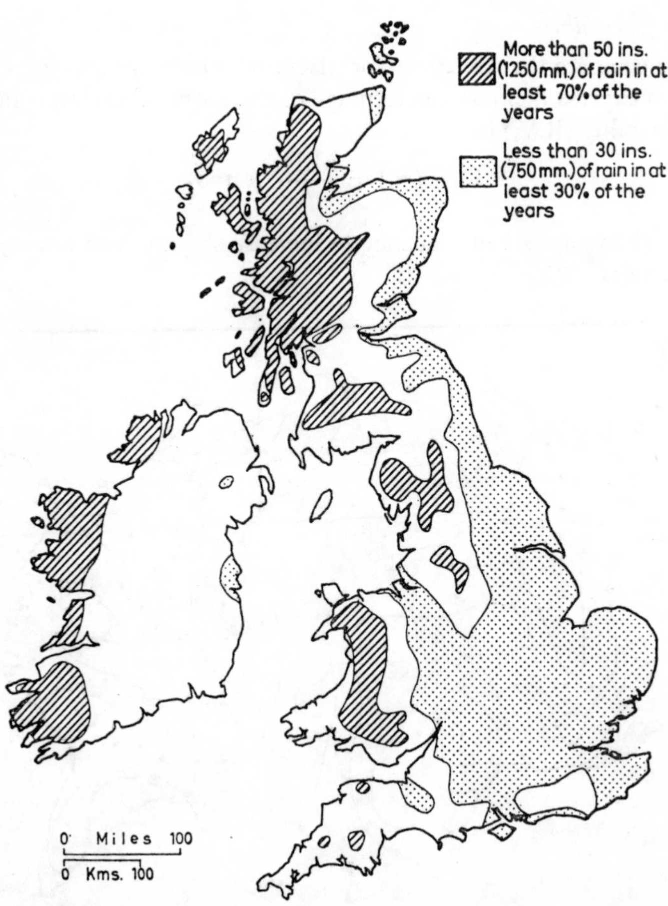

Fig. 3.1(i).4 Areas of reliably high and occasionally low annual rainfall in the British Isles (From Gregory, 1964).
The unshaded areas receive between 30 and 50 in. in at least 70% of years.

D. Probability

If the frequency distribution for a series of annual precipitation totals is symmetrically distributed about the mean and the latter is more or less coincident with the median (i.e. a Normal or Gaussian distribution), probabilities of the occurrence of a specified amount can be determined. The method is set out in most statistical texts; see especially Gregory [1962].

The method has been applied to annual precipitation in several areas of the world; for example, East Africa (Glover *et al.*, 1954) and Great Britain (Gregory, 1957) as illustrated in fig. 3.I(i).4. This shows that large areas of eastern England have at least a 30% probability of receiving less than 75 cm (30 in.), while most upland areas and western Ireland can expect more than 125 cm (50 in.) in at least 70% of years. Such maps are of major significance to agriculturalists and in the assessment of water resources.

6. Dryness and wetness

Abnormal amounts for one part of the world may be quite normal in another. The official British definition of 'absolute drought' is a period of 15 or more consecutive days with each day receiving less than 0·01 in. of rain; a 'wet day' is one with ⩾0·04 in. The former would be quite inappropriate in areas with a long dry season, while the latter would be equally unsatisfactory in the humid tropics. More useful drought definitions involve assessment of effective precipitation and the moisture balance (see Chapter 4.I).

In many parts of the world there seems to be a tendency for dry weather to occur in spells, while the occurrence of a wet day is often independent of the previous conditions. Lawrence [1957] finds that in southern England a dry spell has an increasing probability of continuing (positive persistence) up to about 10 days, whereas after about 30 dry days there is definite likelihood of change (negative persistence). At Tel Aviv an interesting pattern is observed with wet days, ⩾0·1 mm precipitation, in winter. A wet day has a 66% probability of succeeding a wet day, but there is no significant change in the probability if the preceding two or three days were also wet. The occurrence of rain at Tel Aviv is virtually independent of conditions two or more days earlier. This pattern can be described, though not explained, by the statistical model known as a Markov chain. Weiss [1964] shows that the Markov chain model may have wide application to sequences of both wet and dry days in such diverse climatic locations as San Francisco, Moncton (New Brunswick), and Harpenden (England).

The spatial extent of wet and dry extremes on the annual time-scale is also of considerable interest. Glasspoole [1926] analysed the records at 250 stations in the British Isles for 1868–1924 and showed that while 1872 was the wettest for 49% and 1887 the driest for 40% of the whole region, 46 of the 57 years were the wettest or driest of the series *somewhere* in the British Isles. This illustrates the considerable spatial variability of precipitation, even in a zone of predominantly depression control. Spatial variability in areas of low precipitation is commonly underestimated because of the sparse network of rainfall stations.

7. Precipitation characteristics in different macroclimates

Only the briefest sketch of the global variability of precipitation characteristics is possible. We may identify the following properties as being of interest:

annual totals (see Chapter 1.1.4) and their variability;
annual regime;
diurnal regime;
frequency and intensity characteristics.

A. Annual totals and the annual regime

These are the primary climatological characteristics of precipitation. Figure 3.1(i).5 illustrates the theoretical distribution of annual precipitation and its seasonal concentration on a hypothetical continent of low, uniform relief according to Thornthwaite. The actual distribution of summer and winter precipitation is shown in fig. 1.1.4. Six types of regimes are commonly distinguished, although departures from these patterns are numerous. Moreover, the existence of similar regimes does not imply that the mechanisms causing the precipitation are necessarily the same. The six types, illustrated for representative stations in fig. 3.1(i).6, are:

1. *Equatorial.* Rain throughout the year, with two maxima at the equinoxes; generally large annual totals, 250–300 cm or more, and small variability from year to year. Rainfall is associated with the equatorial low-pressure trough, although rainy periods are generally the result of some form of perturbation. In East Africa, for example, outbreaks of rain occur in irregular spells lasting several days (Johnson, 1962). The daily rainfall distribution is determined by large-scale patterns of airflow in the troposphere. The equatorial regime is absent over much of the Indonesian region and in South America other than the Pacific coast.

   Poleward the two equinoctial maxima come closer together, creating a winter dry season. In some areas annual totals are again large, although this regime also occurs on the equatorward margins of tropical deserts.
2. *Tropical* (including monsoon areas). Pronounced summer maximum and winter dry season; annual totals range between about 25 and 100 cm (10 and 40 in.) in the savanna areas where the dry season may last for more than six months, to 200 cm (80 in.) or more in the humid tropics. The regime also extends into the subtropics in eastern Asia. The rainfall in southern Asia and West Africa occurs mainly with disturbances in the monsoon flow south of the equatorial low-pressure trough. There are both convectional downpours and periods of steady rain. In many tropical and subtropical areas late-summer hurricanes make significant contributions to the rainfall.
3. '*Mediterranean.*' Pronounced maximum in the winter half of the year and a dry summer; moderate annual totals of the order of 60–75 cm. It occurs in west-coast subtropical areas. Most of the rainfall is associated with depressions in the westerlies.

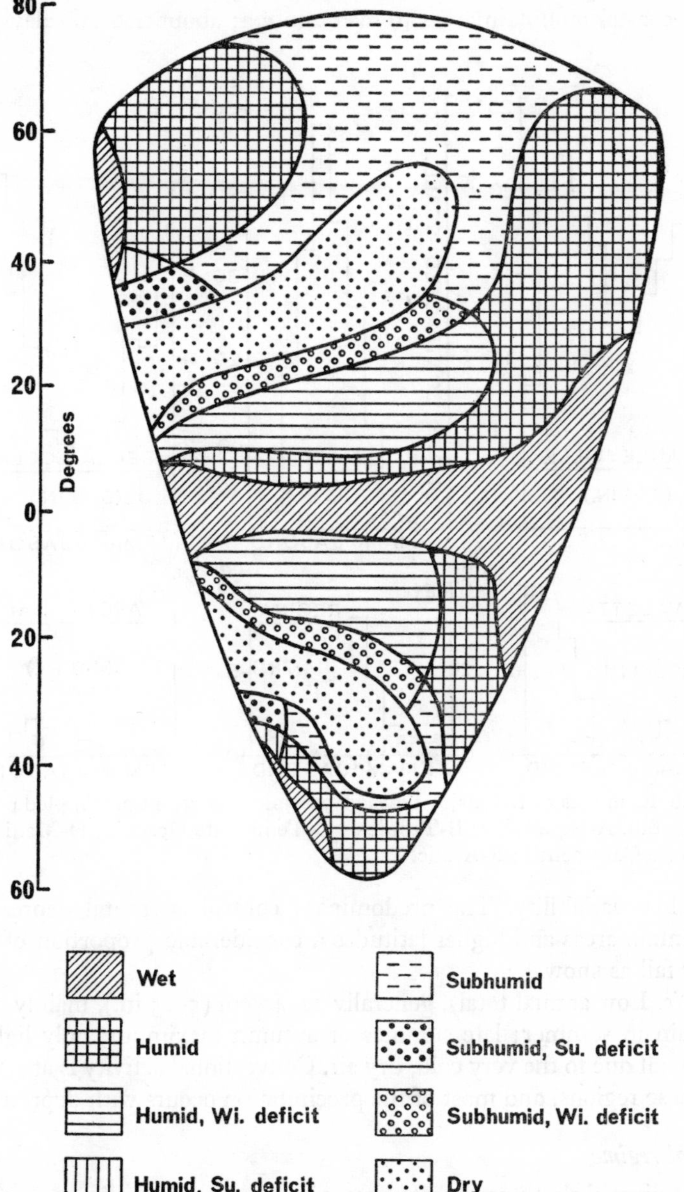

Fig. 3.1(i).5 The hypothetical distribution of precipitation regimes and annual totals on a continent of low, uniform elevation (After Thornthwaite).

4. *Temperate continental interior*. Annual totals of 35–50 cm, mainly occurring as convective rain showers in spring and summer; light winter snowfall; considerable year-to-year variability.
5. *Temperate oceanic* (*west coast*). Precipitation all year with a maximum in winter or autumn. Moderately high totals (75–100 cm) increasing markedly over coastal mountain areas to over 200 cm; about 200 rain days per year

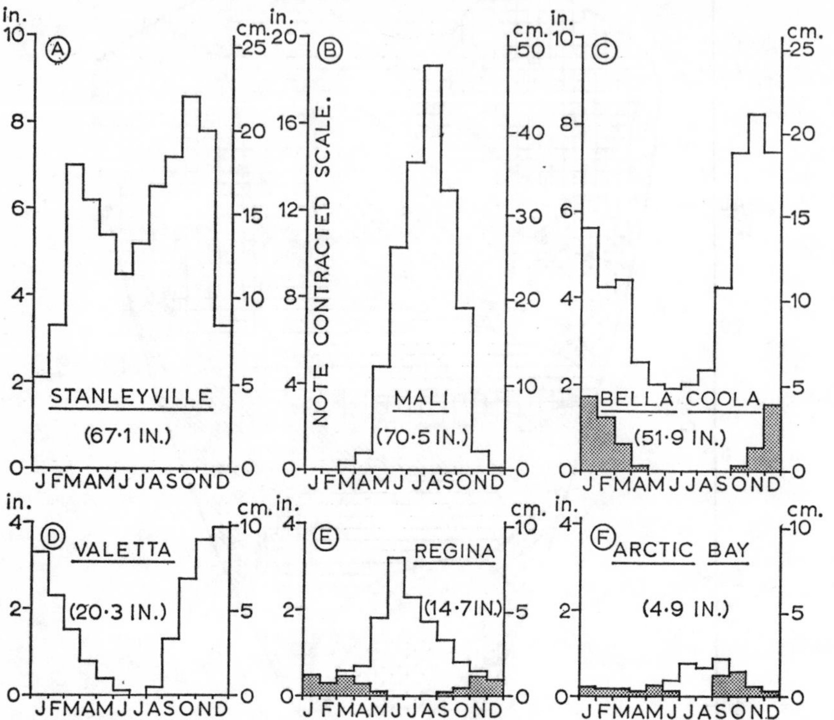

Fig. 3.1(i).6 Examples of the major types of precipitation regimes. Stippled portions indicate Snowfall. A-Equatorial; B-Tropical; C-Temperate Oceanic; D-Mediterranean; E-Temperate Continental; F-Arctic.

and low variability. The predominant control is frontal depressions. In mountain areas and higher latitudes a considerable proportion of the total may fall as snow.

6. *Arctic*. Low annual totals, generally 12–40 cm (5–15 in.), mainly occurring as rain in summer; late summer or autumn maximum; only light winter snowfall due to the very cold, dry air. Convectional activity is at a minimum in these regions, and most of the precipitation occurs with depressions.

B. Diurnal regime

Important diurnal rhythms are often superimposed on the basic seasonal pattern. For example, diurnal heating may lead to an afternoon maximum of precipitation as a result of convective downpours. This regime is popularly regarded as characteristic of tropical climates, but many instances of markedly different diurnal variations are known. Over the tropical oceans a nocturnal maximum is common, while some stations have different regimes in different seasons, so that the diurnal pattern for the year as a whole is indeterminate. At Guam, a small island in the west Pacific (13° N, 145° E), the light showers of the 'dry season' occur mainly between 2200 and 1000 hours, whereas in the 'wet season' there is no clear diurnal rhythm. The afternoon maximum seems to be typical of large islands (of the order of 2,500 ml^2), where convergent sea-breezes initiate deep cumulus leading to heavy showers. It also occurs away from coastal areas, especially where mountains cause uplift of the sea-breezes. This is the case on the Pacific slopes of the Colombian cordillera, while the coast zone has a nocturnal maximum, associated with the effects of land breezes. A more complex illustration is the Malacca Straits in summer. There, the nocturnal maximum is due to convection set-off by the convergence of the land-breeze systems of Malaya and Sumatra (Ramage, 1964). A surprising nocturnal maximum is observed in the Sudan (Oliver, 1965). Khartoum receives 77% of its rainfall between 1800 and 0600 hours. This may be caused by upper winds carrying afternoon thunderstorms south-westward from the hills bordering the Red Sea. However, Bleeker and Andre [1951] suggest that the nocturnal maximum of rainfall and thunderstorms over the central United States is related to a large-scale circulation system induced over the plains by the Rocky Mountains. Several explanations have been suggested for the oceanic nocturnal maximum. Instability may develop in the cloud layer due to radiative cooling of the cloud tops. Air–sea interaction is undoubtedly a contributory factor. The sea surface temperature exceeds the air temperature at night, and so the heat transfer to the air is greatest then, reaching a peak near dawn.

C. Frequency and intensity characteristics

The most generally available frequency statistics are limited to records of the number of days with measurable precipitation or 'rain days'. The mean rainfall per rain day is a rough indicator of rainfall intensity in different climates. For example, this increases from about 4 mm (0·15 in.) per rain day in south Australia to 18 mm (0·70 in.) in tropical north Australia. In the equatorial Kenya Highlands the figure is about 8·4 mm (0·33 in.), whereas in arid northern Kenya it reaches 12·2 mm (0·44 in.) per rain day. Representative values for the United States range from about 5 mm (0·20 in.) per rain day over the Prairies and Great Plains to 14–15 mm (0·55–0·60 in.) in Oklahoma–Arkansas. These values are very low compared with 107 mm (4·2 in.) per rain day in June at Cherrapunji, Assam.

The relative raininess of different types of airflow can be assessed in a similar manner by calculating the 'specific precipitation density'. This is the mean rainfall per rain day for a particular type of airflow as a percentage of the mean for all rain days. At Southampton, England, the mean rainfall intensity in January and July is approximately 5 mm (0·2 in.) per rain day. For different airflow types the specific precipitation–density index is given in Table 3.1(i).3:

The light, showery nature of precipitation with Northerly and North-westerly types is quite apparent. There is an unexpected seasonal change in raininess of Westerly and Cyclonic types.

TABLE 3.1(i).3

Airflow type (H. H. Lamb's classification)	*Specific precipitation density* (%) *January*	(1921–50)	*July*
Northerly	41		54
Easterly	105		121
Southerly	123		130
Westerly	129		80
North-westerly	47		81
Cyclonic	108		163
Anticyclonic	34		14

(from Barry, 1967)

A widespread, perhaps even global, characteristic of rainfall is the occurrence of most of the annual total on a few days. Half of the annual precipitation is accounted for by 13% of the rain days in the Kenya Rift Valley, 16% in the basin of the upper Colorado river, and 10–15% in Argentina. At Concord, New Hampshire, 6% of the rain days gave 23% of the total precipitation during 1885–1935. This characteristic seems, therefore, to be independent of the precipitation regime, annual total, and geographical location.

There are many case studies of short-term precipitation intensity, and although it would be premature to attempt a generalized picture for the various climatic regimes, some pointers in this direction can be indicated. Intensities are generally greater in areas of summer rainfall than winter rainfall. For example, daily intensities in the Middle East average only 5 mm (0·2 in.), with a peak of perhaps 12–13 mm once a year in the mountainous areas, whereas at Tucson, Arizona, summer convective storms with an intensity of 25 mm or more per day account for some 25% of the annual precipitation. For the United States, Paulhus and Miller [1964] have mapped the percentage contribution of daily amounts of 0·5 in. (12·7 mm) or more to the average annual precipitation. The figure ranges from 20% in the Great Basin to 90% on the Gulf Coast. The occurrence of rainfalls exceeding 0·5 in. is considered to be a factor in gully erosion potential. At Namulonge, Uganda, storms giving more than 25 mm per day contributed 30% of the annual total during 1950–55, and peak rates in three rainstorms were between 250 and 350 mm (10 and 13·8 in.) per hour. In tropical northern Australia the daily maximum intensity exceeds 100 mm once a year, and on the Queensland coast it may exceed 150 mm. Daily totals of 150 mm or more are expected only once in a century in Britain, and then principally in upland districts of the west. By contrast, investigations in Idaho and coastal British Columbia indicate no relationship between elevation and intensity.

Rather, the increase with elevation of total amount seems to reflect a longer duration of precipitation.

REFERENCES

BARRY, R. G. [1967], The prospect for synoptic climatology: a case study; In Steel, R. W., and Lawton, R., Editors, *Liverpool Essays in Geography* (Longmans, London), pp. 85–106.

BECKINSALE, R. P. [1957], The nature of tropical rainfall; *Tropical Agriculture*, **34**, 76–98.

BERGERON, T. [1960], Problems and methods of rainfall investigation; In *Physics of Precipitation, Geophysical Monograph No. 5* (Washington), pp. 5–30.

BLEEKER, W. and ANDRE, M. J. [1951], On the diurnal variation of precipitation, particularly over the central U.S.A., and its relation to large-scale orographic circulation systems; *Quarterly Journal of the Royal Meteorological Society*, **77**, 260–71.

CHATFIELD, C. [1966], Wet and dry spells; *Weather*, **21**, 308–10.

COOPER, C. F. [1967], Rainfall intensity and elevation in southwestern Idaho; *Water Resources Research*, **3**, 131–7.

FOSTER, E. F. [1949], *Rainfall and Runoff* (Macmillan, New York), 487 p.

GABRIEL, K. R., and NEUMANN, J. [1962], A Markov chain model for daily rainfall occurrence at Tel Aviv; *Quarterly Journal of the Royal Meteorological Society*, **88**, 90–5.

GILMAN, C. S. [1964], Rainfall; In Ven te Chow, Editor, *Handbook of Applied Hydrology* (New York), Section 9.

GLASSPOOLE, J. [1926], The driest and wettest years at individual stations in British Isles; *Quarterly Journal of the Royal Meteorological Society*, **52**, 237–48.

GREGORY, S. [1957], Annual rainfall probability maps of the British Isles; *Quarterly Journal of the Royal Meteorological Society*, **83**, 543–9.

GREGORY, S. [1962], Statistical Methods and the Geographer (Longmans, London), 240 p.

HARROLD, L. L., and DREIBELBIS, F. R. [1958], *Evaluation of Agricultural Hydrology by Monolith Lysimeters*, 1944–55; Technical Bulletin No. 79, United States Department of Agriculture (Washington), 166 p.

HASTENRATH, S. L. [1967], Rainfall distribution and regime in central America; *Archiv für Meteorologie, Geophysik und Bioklimatologie, Ser.* B, **15(3)**, 201–41.

HERSHFIELD, D. M. [1962], A note on the variability of annual precipitation; *Journal of Applied Meteorology*, **1**, 575–8.

JENNINGS, J. N. [1967], Two maps of rainfall intensity in Australia; *Australian Geographer*, **10**, 252–62.

JOHNSON, D. H. [1962], Rain in East Africa; *Quarterly Journal of the Royal Meteorological Society*, **88**, 1–19.

LAWRENCE, E. N. [1957], Estimation of the frequency of 'runs of dry days'; *Meteorological Magazine*, **86**, 257–69 and 301–4.

MASON, B. J. [1962], *Clouds, Rain and Rainmaking* (Cambridge), 145 p.

NAGEL, J. F. [1956], Fog precipitation on Table Mountain; *Quarterly Journal of the Royal Meteorological Society*, **82**, 452–60.

NYBERG, A. and MODEN, H. [1966], The seasonal distribution of precipitation in the area east of Stockholm and the daily distribution in a few selected cases; *Tellus*, **18**, 745–50.

OLASCOAGA, M. J. [1950], Some aspects of Argentine rainfall; *Tellus*, **2**, 312–18.

OLIVER, J. [1965], Evaporation losses and rainfall regime in central and north Sudan; *Weather*, **20**, 58–64.

PAULHUS, J. L. H. [1965], Indian Ocean and Taiwan rainfall set new records; *Monthly Weather Review*, **93**, 331–5.

PAULHUS, J. L. H. and MILLER, J. F. [1964], Average annual precipitation from daily amounts of 0·50 inch or greater; *Monthly Weather Review*, **92**, 181–6.

RAMAGE, C. S. [1964], Diurnal variation of summer rainfall in Malaya; *Journal of Tropical Geography*, **19**, 62–8.

RODDA, J. C. [1967], A country-wide study of intense rainfall for the United Kingdom; *Journal of Hydrology*, **5**, 58–69.

SAWYER, J. S. [1956], The physical and dynamical problems of orographic rainfall; *Weather*, **11**, 375–81.

SHAW, E. M. [1962], An analysis of the origins of precipitation in northern England, 1956–60; *Quarterly Journal of the Royal Meteorological Society*, **88**, 539–47.

SUZUKI, E. [1967], A statistical and climatological study on the rainfall of Japan; *Papers in Meteorology and Geophysics*, **18**, 103–82.

WALKER, E. R. [1961], *A synoptic climatology of parts of the western Cordillera*; Publications in Meteorology No. 35, Arctic Meteorology Research Group, McGill University (Montreal), 218 p.

WALLÉN, C. C. [1955], Some characteristics of precipitation in Mexico, *Geografiska Annaler*, **37**, 51–85.

WEISS, L. L. [1964], Sequences of wet or dry days described by a Markov chain probability model; *Monthly Weather Review*, **92**, 149–76.

YARNELL, D. L. [1935], Rainfall intensity frequency data; *U.S. Department of Agriculture, Miscellaneous Publication 204* (Washington, D.C.).

3.I(ii). The Assessment of Precipitation

JOHN C. RODDA
Institute of Hydrology, Wallingford

Rain, snow, hail, and sleet, together with dew, rime, and similar phenomena, make up the various forms of precipitation. Most water reaches the surface of the earth as rain, but, of course, snow and dew are important in certain regions. However, both snowfall and dew are difficult to determine, and only rainfall is gauged extensively and with any degree of certainty. Rain gauges are extremely varied in design, and their usage differs considerably. Some are little more than

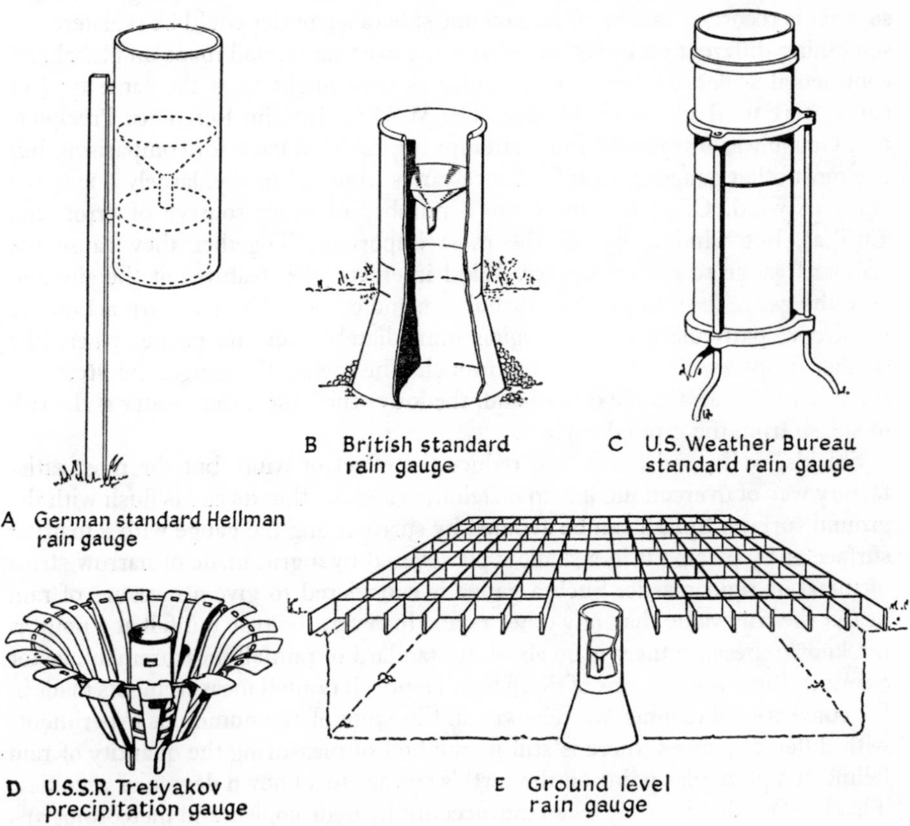

Fig. 3.I(ii).1 Types of standard rain gauges.

a bucket, but others are highly complicated devices that can be interrogated from a distant base for flood-warning purposes. Radar is employed to determine the areal distribution of rainfall, but for quantitative results comparisons are necessary with records obtained from rain gauges.

The history of the rain gauge is lengthy and devious. Some authorities maintain that it commenced in India well over 2,000 years ago. Today there must be hundreds of different types in use, some giving a continuous record of rainfall, but the majority requiring inspection by an observer at a fixed time. However, there is a considerable degree of uniformity in gauge type and observation practice within most national rain-gauge networks – a uniformity which has usually been achieved during the last 100 years. On the other hand, there are appreciable differences from country to country (fig. 3.1(ii).1). At one extreme is the standard gauge of the Soviet Union, standing 2 m high and surrounded by a shield; while at the other is the British standard gauge, a brass cylinder only 1 ft high and 5 in. in diameter. Other national gauges fall between these limits in terms of design and method of installation, but this variety raises the question of how comparable are the results produced by the different gauges. Extensive tests made at the same site show that there are differences from gauge to gauge, so what is recorded as 1 in. of rain on one side of a frontier could be registered as something different on the other. Hence the existing rainfall maps on global and continental scales are not as meaningful as they might be if the same type of gauge were used all the world over. The W.M.O. Interim Reference Precipitation Gauge was introduced in an attempt to provide a basis for comparison, but like most other gauges, its performance varies from site to site, largely due to the effect of wind. Of course there are a number of other sources of error (fig. 3.1(ii).2), but wind is by far the most important. Together they cause the standard gauge to under-register. Wind interacts with features of the site and with the gauge itself to produce turbulence and eddies. These in turn act on the raindrops, particularly in the region immediately over the gauge, where the smaller drops are diverted past the funnel. The higher the gauge, the greater is the effect of wind. On the other hand, the lower the gauge, the greater is the risk of splash from the ground surface.

Shields, walls, and fences can reduce the effect of wind, but the most satisfactory way of overcoming it is to install the gauge so that its rim is flush with the ground surface. Splash can be avoided by surrounding the gauge with a matting surface or by placing it in a shallow pit covered by a grid made of narrow strips of metal or rigid plastic. Such a gauge is considered to give a measure of rain nearer the true value than any other type. However, the true rainfall at a point is not known, because there is no absolute standard of rainfall measurement, as for example, there is in the case of discharge. Hence all rainfall measurements made in the conventional manner are relative, and in spite of the numerous experiments with different gauges, there is still no method of measuring the quantity of rain falling at a particular point on the earth's surface to a known degree of accuracy. This is a fact that is rarely taken into account by hydrologists and meteorologists.

Comparisons of standard gauge observations with those made in nearby

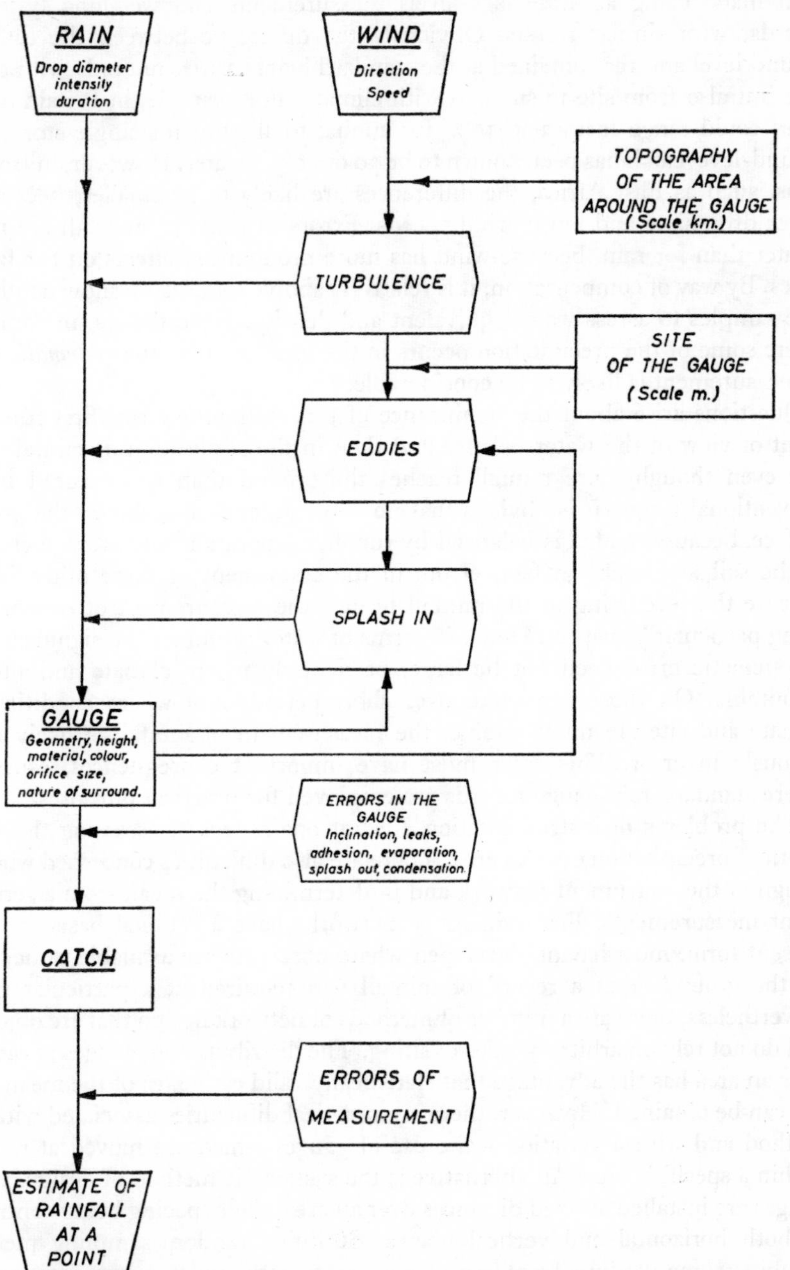

Fig. 3.1(ii).2 Conceptual model of the processes involved in determining rainfall with a conventional rain gauge.

ground-level gauges show that the catch at ground level can be appreciably higher than the catch obtained in the conventional way. Other comparisons have been made using accurate lake-levels measurements and weighing lysimeter records, with similar results. Obviously the difference between the catch at ground level and that obtained at the standard height varies not only with gauge type but also from site to site and with climate. For example, in Britain differences could range from 3 to 10% for annual totals, but for single storms the ground-level catch has been known to be 30 or 40% greater. However, in tropical areas, such as East Africa, the differences are likely to be smaller, because of larger drop sizes and lower wind speeds. Errors in gauging snowfall are much greater than for rain, because wind has more pronounced effects on the falling flakes. By way of compensation, it is relatively simple to measure snow depth and take samples to assess water equivalent and density. Nevertheless, in countries where some of the precipitation occurs in the form of snow the *systematic* error in measurement is likely to be considerable.

Questions arise about the significance of this *systematic* error, first from the point of view of the water balance and then in the application of rainfall data. For even though more rainfall reaches the ground than is measured in the conventional gauge, it is obvious that no extra water is available at the ground surface, because rainfall is balanced by runoff, evaporation, and storage changes in the soil and rock. In fact, errors in the assessment of these other factors obscure that occurring in the rainfall term – the measurement of evaporation being particularly suspect. Hence in terms of water resources the significance of a systematic error need not be large, particularly where climate and site are favourable. On the other hand, over short periods and where conditions of climate and site are unfavourable, the measurement of rainfall is likely to be seriously in error. This error must have important consequences, especially where standard rain-gauge records are employed for practical purposes.

The problems of instrumentation are but one aspect of assessing the mean depth of precipitation over an area. There are also difficulties concerned with the design of the instrument network and in determining the mean from a series of point measurements. Few rain-gauge networks have a rational basis for their present form, most having developed where observers are available rather than on the grounds that a record of rainfall was required at a particular point. Nevertheless, there are a number of methods of network design that are objective and do not rely on arbitrary rules of siting. The distribution of gauges at random over an area has the advantage that statistically valid estimates of the mean rainfall can be obtained. However, there are practical difficulties associated with this method and with a variation – the use of gauges which are moved at random within a specified area. An alternative is the systematic method of design, where gauges are installed at fixed distances over an area, their spacing being controlled in both horizontal and vertical planes. Stratified random-sampling methods combine some of the advantages of both systematic and random design techniques, but there is the difficulty of defining strata in a meaningful way. Of course, the design of any network should be compatible with the use made of the information and with the nature of the topography of the area. For example, a network of gauges installed as a basis for a flood-warning system will be different from one set up to evaluate the water resources of an area, although one set of records could assist in the other objective.

Recording and transmitting gauges, put in a pattern that anticipates the distribution of storm tracks, would produce signals of excessive rates of rainfall and others indicating that some predetermined amount of rain had fallen in a given time. Such a flood-warning system would suffer from the problem of delimiting the area of a storm and the difficulties of estimating intensities at places between the gauges. One solution might be to use a secondary gauge network and build up a history of space-intensity relationships from past storm records. Another would be to provide radar coverage, but this is not often possible because of cost. Non-recording gauges would be employed in the case of a water-resources survey. A wide coverage of the area would be aimed at to provide information on spatial variations, while variations in time could be accounted for by incorporating long-established gauges in the network. It could be argued that networks designed for the same purpose would need to be more dense in mountainous areas than in flat country, because of the wider variations in rainfall that are to be found where differences in relief are most marked. However, this matter is not clear, because similar patterns of rainfall occur over mountainous areas, patterns which appear to be highly correlated with one another; as opposed to the more random distribution of rain that takes place in flat areas.

The transformation of point measurements of rainfall into an estimate of the mean for an area can be carried out in several ways. Where gauges are evenly distributed over an area and relief is subdued use can be made of the arithmetic mean. An advance on this is the construction of 'Thiessen' polygons around each gauge, as then each gauge record is weighted according to the area of the polygon around it. This method is objective by contrast with the isohyetal and isopercentile techniques, which are largely subjective. One other method is the use of regression analysis, but this is only successful in regions where topography controls the distribution of rain.

It is highly probable that the error in determining the mean rainfall for an area will be appreciable, even when the most satisfactory instruments are combined with the best techniques of network design and computation of the mean. Where snow is important this error will be even greater.

REFERENCES

PONCELET, L. [1959], *Sur le comportement des Pluviometres*; Publications, Series A, No. 10 Institut Royal Meteorologique de Belgique.

KURTYKA, J. C. [1953], *Precipitation Measurements Study*; Report of Investigation No. 20 State Water Survey Division, Illinois.

RODDA, J. C. [1967], The Rainfall Measurement Problem; *Proceedings of the Bern Assembly*, International Association of Scientific Hydrology.

WORLD METEOROLOGICAL ORGANIZATION [1965], *Guide to Hydrometeorological Practices*; Bulletin No. 168 t.p. 82.

4.I. Evaporation and Transpiration

R. G. BARRY

Institute of Arctic and Alpine Research, University of Colorado

Investigation of the transfer of moisture from the surface of the earth to the atmosphere concerns workers from a number of disciplines. On the practical side there are agriculturalists, foresters, and hydrologists, and on the theoretical side meteorological physicists and plant physiologists. The physical controls on evaporation have been recognized since 1802, when John Dalton first stated the basic principles, but it is only during the last twenty or so years that the active exchange of practical and theoretical findings between research workers has begun to provide a coherent body of knowledge. Inevitably there are innumerable specialized papers on evaporation reflecting these different approaches, and the treatment here is necessarily restricted to a statement of the basic concepts and an outline of some applications and results.

More detailed accounts of the theory are provided by King [1961], Sellers [1965, pp. 141–80], and Thornthwaite and Hare [1965].

1. Basic mechanisms of evaporation

Net transfer of water molecules into the air occurs only if there is a vapour-pressure gradient between the evaporating surface and the air, i.e. evaporation is nil when the relative humidity of the air is 100%. Evaporation from a moist surface involves a change of state from liquid to vapour, and therefore necessitates a source of latent heat. To evaporate 1 g of water requires 540 cal of heat at 100° C and 600 cal at 0° C. An external heat source must therefore be available. This may be solar radiation, sensible heat from the atmosphere, or from the ground. Alternatively, it may be drawn from the kinetic energy of the water molecules, thus cooling the water until equilibrium with the atmosphere is established and evaporation ceases. In general, solar radiation is the principal energy source for evaporation.

In addition to the two primary controls, the evaporation rate is affected by wind speed, since air movement carries fresh unsaturated air to the evaporating surface. Within approximately 1 mm of the surface the upward movement of vapour is by individual molecules ('molecular diffusion'), but above this surface boundary layer turbulent air motion ('eddy diffusion') is responsible. The temperature of the evaporating surface also affects evaporation. At higher temperatures more water molecules can leave the surface due to their greater kinetic energy. Salinity depresses the evaporation rate in proportion to the solution concentration. For sea-water the rate is about 2–3% lower than for fresh water.

2. Plant factors

Water loss from plants – *transpiration* – takes place when the vapour pressure in the air is less than that in the leaf cells. About 95% of the diurnal water loss occurs during the daytime, because water vapour is transpired through small pores, or *stomata*, in the leaves, which open in response to stimulation by light. Transfer of water vapour to the atmosphere is the initiating process in the movement of water from the soil via the plant. It is a passive process so far as the plant is concerned, but it performs a vital function in effecting the internal transport of nutrients and in cooling leaf surfaces. Transpiration considerably exceeds the direct water needs of the plant. Nevertheless, the transfer of water to the air is unavoidable. In the absence of a plant cover evaporation would still occur from the soil.

Interaction between soil-moisture content and root development is a complicating factor. If soil water is not replenished over a period of weeks vegetation with deeper roots, especially trees, will transpire more than shallow-rooted plants, other things being equal (see Chapter 4.1.8). Some support for this idea is provided by catchment studies. Run-off from catchments under grass generally exceeds that from catchments under woodland. However, this problem remains a subject of considerable controversy.

Resistances to water movement, both in the soil and in plant tissues, must be considered. These include soil-water tension, the resistance of cell walls in the roots and leaves to water transport, and the resistance of stomata to vapour transfer. The *internal* (stomatal) *resistance* of a single leaf to diffusion is an important control on transpiration, and it is dependent on the size and distribution of the stomata. For a crop or vegetation cover with several leaf layers the effective stomatal resistance (r_s) is reduced to approximately 30% of that of an individual leaf, owing to the decreased ventilation within the cover. Seasonal variations associated with changes in the leaf area affect r_s, as do diurnal variations. The latter result partly from the opening and closing of the stomata with light intensity and partly from the effects of transpiration stress on the stomata when water uptake lags behind transpiration. A separate *external resistance* of the air to molecular diffusion (r_a) arises through frictional drag of air over the leaf (larger leaves have lower transpiration rates) and the interference between diffusing molecules of water vapour. A decrease in r_a may be due to higher wind speeds or greater 'roughness' of the vegetation surface, which causes increased turbulence in the air flow. Generally the stomatal resistance r_s is larger than r_a, although, as discussed below (Chapter 4.1.8), the interaction of r_s and r_a is an important determinant of evaporation rates.

A further effect of a vegetation cover is that it intercepts precipitation before it can reach the surface. A forest canopy may retain up to 30% of total precipitation (more for conifers than deciduous species), and the proportion is larger for light, showery precipitation.

The amount reaching the ground via stem flow varies according to tree-type, but the bulk is evaporated without entering into the soil–plant part of the cycle. This might be regarded as an excessive loss of moisture compared with a grass cover, especially in the winter. However, the radiant energy used in evaporating intercepted water is unavailable for other evapotranspiration, and hence interception is not as serious a problem as it might appear to be.

3. Potential evapotranspiration

Moisture transfer from a vegetated surface is often referred to as evapotranspiration,[1] and when the moisture supply in the soil is unlimited the term potential

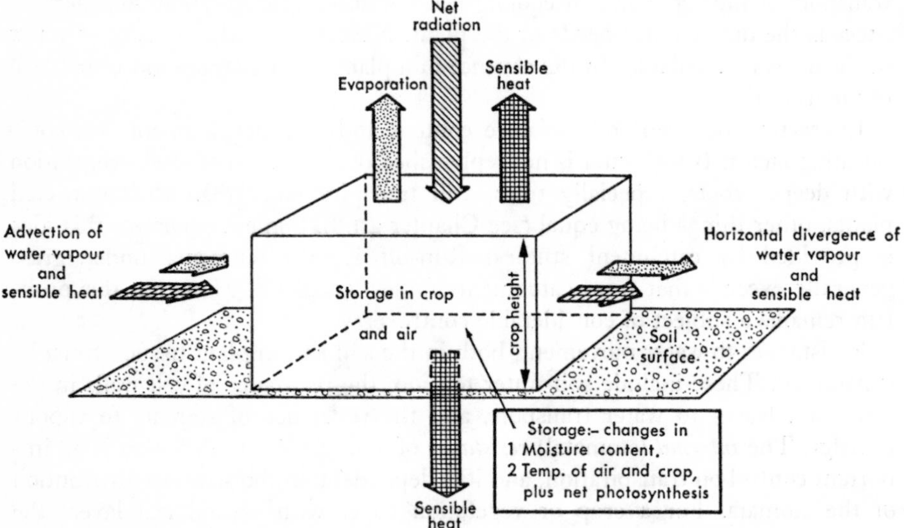

Fig. 4.I.I The energy balance at the surface (After King, 1961).

evapotranspiration (PE) is used. It has been suggested that PE can be defined more specifically as the evaporation equivalent of the available net radiation, i.e. $PE = R_N/L$, where L is the latent heat of vaporization (59 cal $cm^{-2} \approx 1$ mm evaporation). In some cases this equivalence may be invalid. For example, if an irrigated area is surrounded by dry fields evaporation rates can exceed R_N/L by 25–30%. Air heated by passing over the dry areas upwind maintains the high rates through the downward transfer of sensible heat to the irrigated section – the so-called 'oasis effect'. Horizontal transport (advection) of sensible heat *through* the vegetation cover (fig. 4.I.I) can also cause anomalous evaporation rates – termed the 'clothesline effect'. This occurs when a study plot is not

[1] In agricultural studies the term 'consumptive use' (*CU*) of water by crops is commonly used. However, in irrigation engineering, where the term is applied to irrigated crops, $CU = PE$. At certain times of year CU is less than PE. Consequently, there is a risk of confusion and misinterpretation.

surrounded by a zone with identical vegetation cover and environmental conditions. The 'buffer zone' necessary to eliminate these effects varies in size, but may exceed 300 m radius. Neverthless, for all short crops of approximately the same colour and completely covering the ground the *PE* rate is essentially determined by the total available energy as long as there is unlimited soil water. Plant physiology is important in the case of specialized crops, such as rice and sugar cane (high water use rates) and pineapple (low usage).

4. Actual evaporation

It is known that when the moisture supply in the soil is limited plants have difficulty in extracting water, and the evaporation rate (*E*) falls short of its maximum value (*PE*). The precise nature of this relationship is controversial. One view is that the potential rate is maintained until soil-moisture content

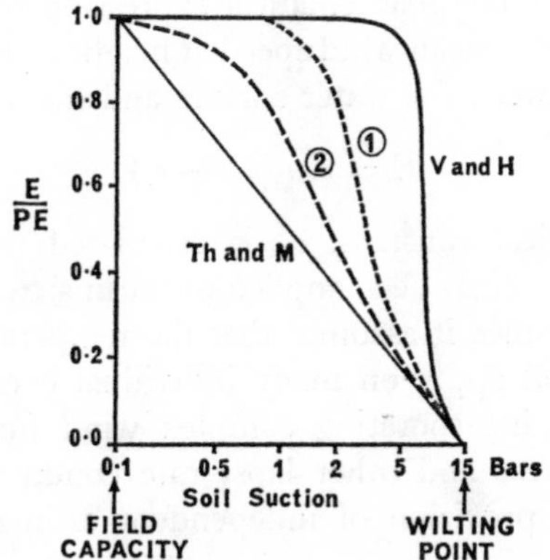

Fig. 4.1.2 The relationship between the ratio of actual to potential evapotranspiration $\left(\frac{E}{PE}\right)$ and soil moisture (After Holmes, 1961, and Chang, 1965).

V and H = Veihmeyer and Hendrickson.
Th and M = Thornthwaite and Mather.
1 and 2 = Schematic curves for a vegetation-covered clay-loam under low evaporation stress and a vegetation-covered sandy soil under high evaporation stress, respectively.
1 Bar = 1,000 millibars (10^6 dynes/cm^2).

drops below some critical value, after which there is a sharp decrease in evaporation, while another is that the rate decreases progressively with diminishing soil moisture. At field capacity (maximum soil moisture content under free drainage) $E/PE = 1$, i.e. evaporation proceeds at the maximum potential rate. Veihmeyer and Hendrickson consider that no change takes place in this ratio until the plant is near wilting point (fig. 4.1.2). Thornthwaite and Mather assume the decrease below field capacity to be a logarithmic function of soil suction, but recent work suggests that $E/PE \approx 1$ as long as the moisture content is at least 75% of field capacity. Undoubtedly the soil type and climatic conditions are important; field capacity ranges from 25 mm in a shallow sandy soil to 550 mm in deep clay-loams. Chang [1965] indicates that Veihmeyer and Hendrickson's results may apply to a heavy soil with vegetation cover in a humid, cloudy region, whereas in sandy soils with a vegetation cover under arid conditions a rapid decline in E/PE is likely. Experimental work by Holmes [1961] supports this view; see lines (1) and (2) on fig. 4.1.2.

5. Meteorological formulae

There are two principal lines of approach to estimating evaporation through physical relationships; one is the aerodynamic (or mass transfer) method, the other is the energy budget method.

A. Aerodynamic method

This method considers factors controlling the removal of vapour from the evaporating surface. These are the vertical gradient of humidity and the turbulence of the air flow. The mathematical expression relates evaporation from (large) water bodies to the mean wind speed at height z (u_z), and the mean vapour pressure difference between the water surface and the air at level z ($e_w - e_z$),

$$E = Ku_z(e_w - e_z) \qquad (1)$$

where K is an empirical constant. e_w is calculated for mean water surface temperature. The method has been applied to ocean areas in particular, but only for monthly averages, since it assumes that the temperature lapse rate is adiabatic, and this does not apply on many individual occasions. More elaborate forms of equation (1) incorporating complex wind functions have been developed for land surfaces and other lapse-rate conditions, but their value is limited mainly to the provision of independent estimates of evaporation for research purposes.

B. Energy budget method

From fundamental principles of the conservation of energy it follows that the net total of long- and short-wave radiation received at the surface (R_N) is available for three processes (fig. 4.1.1): the transfer of sensible heat (H) and of latent heat (LE) to the atmosphere and of sensible heat into the ground (G). That is,

$$R_N = H + LE + G \qquad (2)$$

The fraction of R_N used in plant photosynthesis is generally negligible. Accordingly, evaporation can be determined by measurement of the other terms

$$E = \frac{R_N - H - G}{L} \qquad (3)$$

R_N can be measured by the use of a net radiometer, and G is calculated from data on the soil-temperature profile or by direct measurement of soil heat flux, but H cannot readily be estimated. An indirect method is to employ Bowen's ratio $\beta = H/LE$. This is calculated from the ratio of the vertical gradients of temperature and vapour pressure. However, the determination is unreliable when the surface is dry and H is large. On substitution of β in (3)

$$E = \frac{R_N - G}{L(1 + \beta)} \tag{4}$$

The use of Bowen's ratio assumes that the vertical transfer of heat and water vapour by turbulence takes place with equal efficiency. Recent work in Australia (Dyer, 1967) shows that this assumption is universally valid. Given the requisite observational data, the energy budget approach is a practicable one for determining evaporation over periods as short as an hour.

C. Combination methods

A number of methods have been developed to combine the aerodynamic and energy budget approaches, thereby eliminating certain measurement difficulties which each presents. The most widely used combination method was derived by Penman [1963, p. 40]. He expresses PE as a function of available radiant energy (R_N) and a term (E_a) combining saturation deficit and wind speed.

$$R_N = 0{\cdot}75S - L_N \tag{5}$$

where $0{\cdot}75S$ = solar radiation absorbed by a grass surface;

L_N = net long-wave (terrestrial) radiation from the surface.

$$E_a = f(u)(e_s - e) \tag{6}$$

where $f(u) = 0{\cdot}35\,(1 + 0{\cdot}01u)$ for short grass;

u = wind speed at 2 metres (miles/day);

e_s = saturation vapour pressure (mm mercury) at mean air temperature;

e = actual vapour pressure at mean air temperature and humidity.

The expression for PE from short grass[1] is

$$PE\ (\text{mm/day}) = \frac{\left(\frac{\Delta}{\gamma}\frac{R_N}{L} + E_a\right)}{\frac{\Delta}{\gamma} + 1} \tag{7}$$

where $\frac{\Delta}{\gamma}$ = Bowen's ratio;

$\gamma = 0{\cdot}27$ (mm mercury/° F), the psychrometric constant;

$\Delta = \frac{de_s}{dt}$, the change of saturation vapour pressure with mean air temperature (mm mercury/° F).

It is worth while examining certain of these terms further. In equation (5) the 0·75 weighting of the incoming solar radiation is due to the 25% albedo (re-

[1] In Penman's original formulation evaporation was determined first for an open-water surface (PE_0), then a weighting factor (f) of 0·6–0·8 was used according to season and type of surface. $PE = f\,.\,PE_0$.

flection coefficient) of short grass. Most green crops have a similar reflectivity, but values for coniferous forest and heath are approximately 15%, while an average value for water is 5%. This factor should augment evaporation from a water surface compared with grass or crops, but at least a partial compensation is provided by the greater aerodynamic roughness of these surfaces (see Chapter 4.1.8). Rutter, for example, finds an annual evaporation of 679 mm from Scots Pine (*Pinus sylvestris*) in Berkshire for 1957–63, compared with a calculated open-water evaporation of 597 mm. Observational evidence is by no means unanimous, however, on this point. The meaning of the saturation deficit term

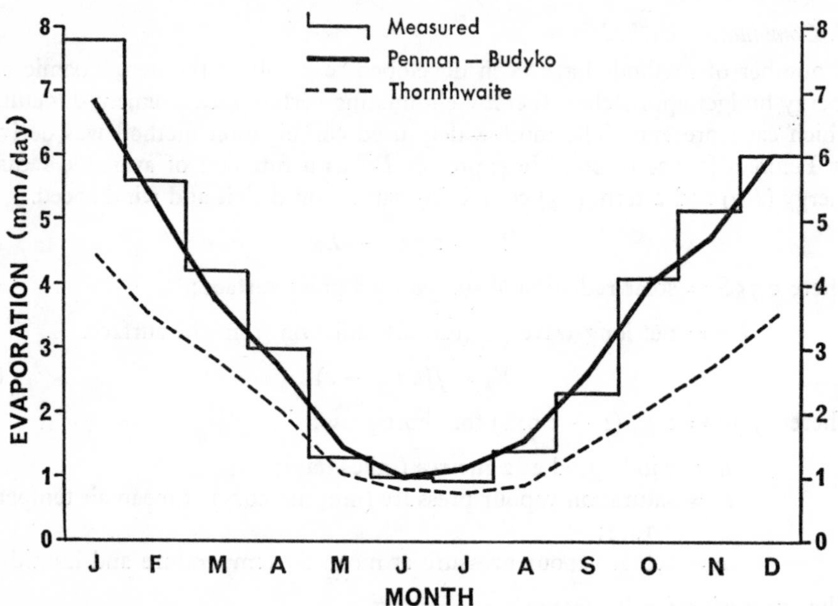

Fig. 4.1.3 Lysimetric observations at Aspendale, Australia (38° S), compared with estimates from the Penman–Budyko and Thornthwaite methods (Based on data from McIlroy and Angus, and Sellers, 1965).

$(e_s - e)$ in equation (6) is a common source of misunderstanding. It represents the 'drying power' of the air, but this need not be directly related to evaporation. In fact, $(e_s - e)$ is likely to be greatest when the surface is very dry and no moisture is available for evaporation. Certain aspects of this approach require further comment. First, the basis of the formulation involves the assumptions of Bowen's ratio, discussed in the previous section. Second, the transfer of sensible heat into the ground is neglected. Third, the use of mean temperatures and humidities makes it unsuitable for short-period estimates (<24 hours) of evaporation rates.

Budyko independently derived a similar expression for *PE*, and fig. 4.1.3 illustrates the accuracy of the Penman–Budyko approach compared with lysimetric observations in Australia.

A much simplified approach incorporating saturation deficit and radiation has been developed by Olivier. The equation for PE (mm/day) is

$$PE = (T - T_w)\,\frac{\bar{L}}{L^2} \tag{8}$$

where $L = \frac{S}{S_v}$, $\bar{L} = \frac{\bar{S}}{\bar{S}_v}$;

S = total solar radiation under clear skies for the latitude of the station for a particular month;

S_v = vertical component of S;

$\bar{S}$ = average of the 12 monthly values of S;

$\bar{S}_v$ = average of the 12 monthly values of S_v;

T = mean monthly temperature (° C);

T_w = mean monthly wet-bulb temperature.

Figure 4.1.4 shows the estimate of PE obtained in three different regimes using Penman's and Olivier's formulae.

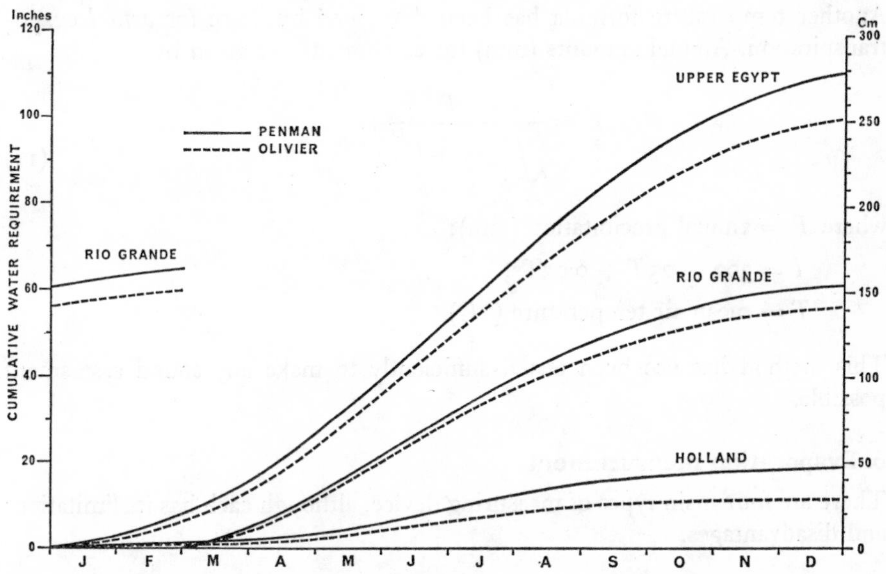

Fig. 4.1.4 Estimates of potential evapotranspiration in three different climatic regimes using Penman's and Olivier's methods (After Olivier, 1961).

D. Temperature formulae

One of the best-known methods of estimating PE was developed by Thornthwaite. He related observations of consumptive use of water in irrigated areas in the western United States to air temperatures, with adjustments for daylength.

$$PE\ \text{(mm/month)} = 16\left(\frac{10T}{I}\right)^a \tag{9}$$

where T = mean monthly temperature (° C);

a = an empirical function of I;

$$I = \sum_{1}^{12} \left(\frac{T}{5}\right)^{1.514}.$$

The values can be readily calculated from published tables or nomograms. The method has been widely applied, although in some climatic regimes it gives unreliable results, as fig. 4.1.3 indicates for south-east Australia.

A more soundly based relationship has been illustrated by Budyko. He shows that if heat storage and sensible heat transfer are effectively zero annual *PE* is given approximately by

$$PE \text{ (mm/year)} \approx \frac{R_{N0}}{L} \approx 0{\cdot}18 \sum T \tag{10}$$

where R_{N0} = the net radiation budget of a wet ground surface;

$\sum T$ = the sum of daily mean temperature which exceed 10° C.

Another temperature formula has been developed by Turc for *actual* evapotranspiration. Annual amounts (mm) for catchments are given by

$$E = \frac{P}{\sqrt{\left\{0{\cdot}9 + \left(\frac{P}{I}\right)^2\right\}}} \tag{11}$$

where P = annual precipitation (mm);

$I = 300 + 25T + 0{\cdot}05T^3$;

T = mean air temperature (° C).

This method has not been tested sufficiently to make any sound assessment possible.

6. Evaporation measurement

There are four main types of measuring device, although each has its limitations and disadvantages.

A. Atmometers

These are water-filled glass tubes having an open end through which water evaporates from a filter-paper (Piche type) or porous plate (Bellani type). The tube supplying water is graduated to read evaporation in mm, but the evaporation (termed 'latent evaporation') can only be compared with readings from another such instrument in a similar exposure, and may bear little or no relation to evaporation from land or water surfaces, since it only reflects the saturation deficit of the air (see p. 88). The instrument is apparently more responsive to wind speed than radiant energy.

B. Evaporation pans

The 'Class A' pan of the United States Weather Bureau, which is approved by the World Meteorological Organization, is 122 cm (48 in.) in diameter and 25 cm (10 in.) deep. Problems can arise through splashing, heating of the pan walls, and interference by birds or animals, while the installation position (sunken or mounted on or above the surface) is particularly critical. Unfortunately, pan evaporation (E_p) is not related to lake evaporation (E_L) in any simple or constant manner. Kohler found that E_L/E_p is generally within the range 0·6–0·8 for the United States.

In general, evaporation decreases as the size of the water body increases. In part, this arises from the 'oasis' effect. Air travelling over a large water surface picks up sufficient moisture to reduce the evaporation rate towards the leeward shore. Water depth is another cause of pan–lake differences. Much energy in spring goes into heating a deep lake, thereby suppressing evaporation rates. Morton calculates the average annual evaporation from Lake Ontario as 813 mm (32·0 in.), whereas that from Lake Superior is only 546 mm (21·5 in.). In the case of pan–lake comparisons these seasonal effects are even more serious, and it is only safe to use pan data for annual estimates of lake evaporation.

Even allowing for the regional and other factors which affect the ratio E_L/E_p (the 'pan coefficient'), the measurements provide no indication of evaporation from a land surface. The other two approaches now to be described are directed to this end.

C. Lysimeters

A lysimeter is an enclosed block of soil with a vegetation cover (usually short grass) similar to that of its surroundings. Figure 4.1.5 illustrates the installation at Hancock, Wisconsin. At regular intervals the weight change (ΔS), precipitation (P), and percolation (r) are measured. By means of the moisture-balance equation, evapotranspiration is determined as a residual.

$$E = P + \Delta S - r \tag{12}$$

The use of a large block allows an accuracy of 0·01 in. of water-depth.

A simpler version for PE determination is the 'Thornthwaite' type of evaporimeter. Here the moisture supply is maintained by 'irrigating' the block when necessary, so that the soil-moisture storage can be regarded as constant. The percolation (r), precipitation (P), and added water (W) are measured.

$$PE = P + W - r \tag{13}$$

With both types of device the presence of the tank's base may interfere with the soil moisture profile compared with that in a natural soil unless special precautions are taken.

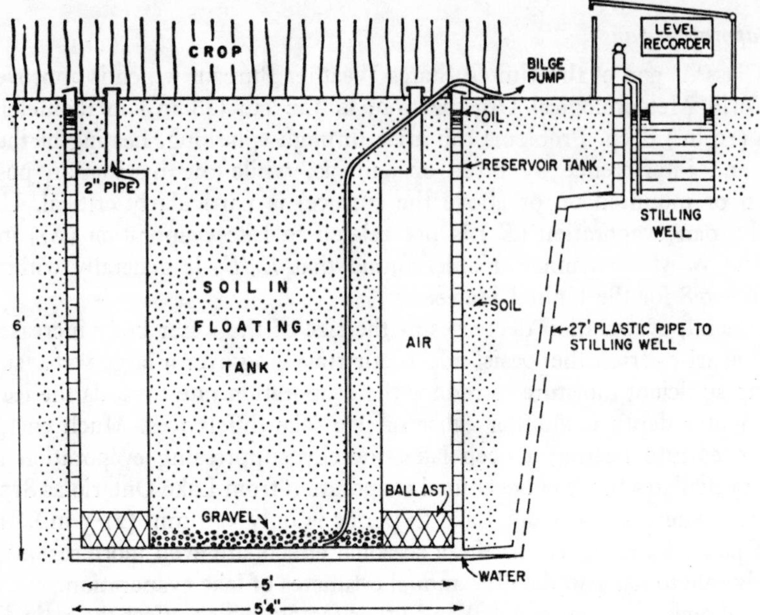

Fig. 4.1.5 Lysimeter installation at Hancock, Wisconsin (From King, 1961).
Here the soil block floats in a tank of water. Changes of water level are recorded instead of weighing the block.

D. The 'evapotron'

Attempts have recently been made to measure the vertical transfer of moisture directly. Instruments developed by C.S.I.R.O. in Australia measure the magnitude and direction of vertical eddies which transfer water vapour upwards. There are many difficulties with this approach. In particular, there is a need for instruments that measure instantaneous changes of both the vapour content and vertical velocity of the air. The subsequent determination of average evaporation rates requires a computer to integrate the results. Moreover, effects of advection and storage below the measuring level (see fig. 4.1.1) may present serious difficulties unless measurements are made near the surface. This technique seems likely to be limited to research applications.

7. Budget estimates

The moisture-budget equation already referred to can be used in two very different ways to estimate evapotranspiration from large areas over a time period of the order of a month.

A. Catchment estimates

If suitable allowance can be made for storage in the catchment system, or if it is assumed constant, over a sufficiently long time interval,

$$E = r - P \tag{14}$$

where r is the runoff measured by river gauging. Checks of this kind showed that estimates of evaporation from the Thornthwaite formula in northern Finland and northern Labrador–Ungava were 80% too high. This may reflect low snowfall estimates as well as low water use by moss and lichen surfaces.

B. Aerological estimates

In analogous manner evaporation can be estimated from data on atmospheric moisture (see Chapter 1.1.5).

$$E = \Delta D - P \pm \Delta S \tag{15}$$

where ΔS is the storage change in the overlying air column and ΔD is the net divergence (or convergence) of water vapour out of (or into) the column. This method requires very complete aerological records.

8. Evaporation rates in different macroclimates

The global pattern of annual evaporation has been outlined in Chapter 1.1 and we can now look in more detail at variations in seasonal regime.

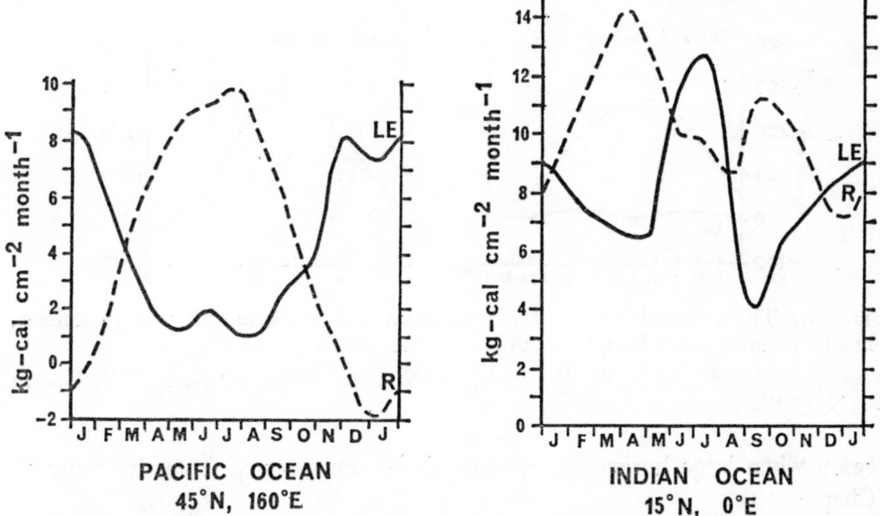

Fig. 4.1.6 The seasonal march of net radiation and evapotranspiration over ocean areas (After Budyko, 1956).

For the ocean areas of the world an average of 90% of annual R_N is used for evaporation, whereas for the continents the figure is only just over 50% and the remainder represents sensible heat transferred to the soil and the atmosphere. The inverse correlation of the seasonal march of R_N and LE over the oceans is therefore at first sight unexpected (fig. 4.1.6). This is a result of the complex interaction of heat storage and heat transfer by ocean currents. Much of the required energy for ocean evaporation is derived from the water itself, and the rate is mainly determined by wind speed and the vapour-pressure gradient. Over the Indian Ocean the summer maximum is caused by the higher wind speeds, while cloudiness diminishes the radiation receipt. The secondary winter maximum is due to the advection of dry trade-wind air. Off the eastern shores of Asia and North America there are large evaporation losses in winter as cold, dry continental polar air flows across the warm Gulf Stream and Kuro Shio Currents. In summer, however, reduced wind speeds and low air–sea vapour-pressure differences suppress evaporation rates. It may be recalled that this is the

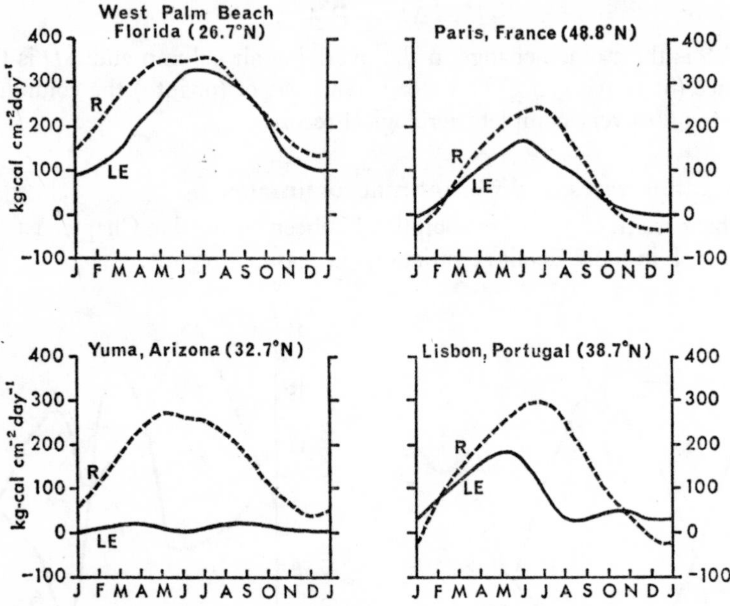

Fig. 4.1.7 The seasonal march of net radiation and evapotranspiration in different climatic regimes (After Budyko, 1956 and Sellers, 1965).

West Palm Beach, 27° N, 80° W; Paris, 49° N, 2° E; Yuma, 33° N, 115° W; Lisbon, 39° N, 9° W.

season when large horizontal moisture fluxes are directed from the continent (Chapter 1.1).

Over land, the seasonal regime generally reflects the occurrence of maximum net radiation receipts and maximum surface-air vapour pressure difference. Where precipitation occurs mainly in summer, or has an even distribution throughout the year, there is a simple summer maximum and winter minimum of evaporation. Figure 4.1.7 illustrates typical profiles for West Palm Beach, Florida, and Paris. In areas of summer drought and winter rains there is a spring evaporation maximum, such as at Lisbon, while in districts with rains in autumn and winter there may be a double maximum in spring and autumn as at Yuma, Arizona.

Regional and local differences in evaporation rate arise not only from variations in the meteorological controls but also from largely independent soil and vegetation factors. For example, observations in the Canadian Subarctic indicate that $LE/R_N \approx \frac{1}{3}$ over lichen (*Cladonia* spp.) surfaces. The low evaporation rate is apparently due to the negligible extraction of moisture from the soil by non-vascular vegetation. Local differences may also reflect the varying external (r_a) and internal (r_s) resistances of vegetation surfaces to vapour diffusion (see 4.1.2). Theoretical computations by Monteith indicate a potential transpiration in the Thames valley, England, of 47 cm/year for short grass, compared with 58 cm/year for a tall farm crop with smaller r_a due to greater surface roughness.

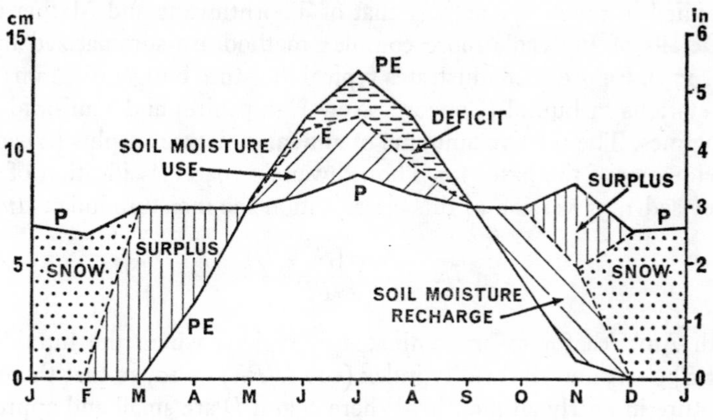

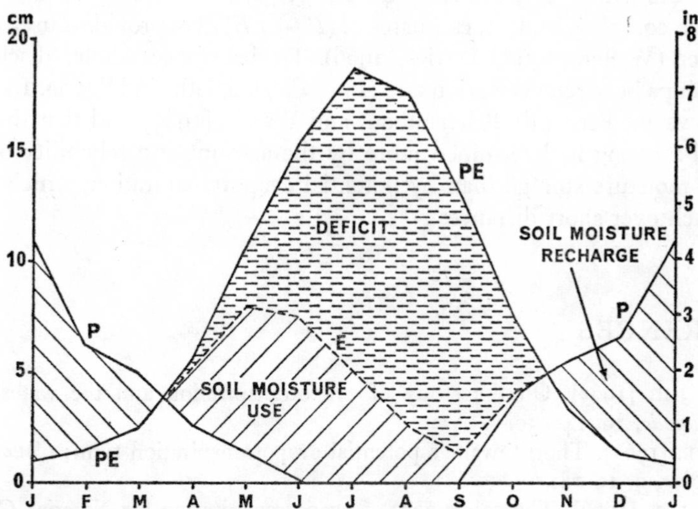

Fig. 4.1.8 Moisture budget diagrams for Concord, New Hampshire (*above*), and Aleppo, Syria (*below*) (Based on data in Thornthwaite and Hare, 1965, and Mather, 1963).

The method assumes that 50% of the soil water surplus runs off in the first month, 50% of the remainder in the next, and so on, unless additional surplus forms.

The loss from a pine forest (48 cm/year) is almost the same as from grass because increased stomatal resistance offsets the influence of lower albedo (15% compared with 25% for grass and green crops) and greater surface roughness of the forest which otherwise tend to promote transpiration.

9. The moisture balance and some applications

The principal difficulty encountered in computing moisture budgets for individual localities is the problem of assessing soil-moisture storage and actual evaporation. Only the simplest of the models outlined in 4.1.4 has been extensively applied in practice – namely that of Thornthwaite and Mather (Mather, 1963). Details of Budyko's more complex method are summarized by Sellers [1965, p. 175]. Figure 4.1.8 illustrates typical moisture budget diagrams for mid-latitude stations in humid (Concord, New Hampshire) and semi-arid (Aleppo, Syria) regimes. The relative amounts of annual moisture surplus (S) and deficit (D) provide one of the bases for Thornthwaite's 1948 classification of climates. In the revised 1955 version of this classification the moisture index (Im) is

$$Im = \frac{100\,(S - D)}{PE} \tag{16}$$

In North America forest predominates in regions where the humidity index ($100\,S/PE$) > 35 and the aridity index ($100\,D/PE$) < 10, so that there is ample soil moisture in nearly all months. Where S and D are small and approximately equal, the vegetation is typically tall grass prairie, but where the aridity index is of the order of 30 this gives way to short grass. Sagebrush and other desert vegetation occurs with aridity indices >40. For regions where data are inadequate to calculate a complete budget, estimates of ($P - PE$) may provide a useful climate parameter (Wallén, 1966; Davies, 1966). Davies, for example, discusses the relationships between vegetation and ($P - PE$) isopleths in Nigeria. In the Near East, however, Perrin de Brichambaut and Wallén [1963] find that the limit of dry-land farming is determined more by the amount and reliability of rainfall and soil-moisture storage than by potential evapotranspiration, which does not vary much over short distances.

REFERENCES

CHANG, J-H. [1965], On the study of evapotranspiration and the water balance; *Erdkunde*, **19**, 141–50.

CURRY, L. [1965], Thornthwaite's potential evapotranspiration term; *Canadian Geographer*, **9**, 13–18.

DAVIES, J. A. [1966], The assessment of evapotranspiration for Nigeria; *Geografiska Annaler*, **48**, Ser. A., 139–56.

DYER, A. J. [1967], The turbulent transport of heat and water vapour in an unstable atmosphere; *Quarterly Journal of the Royal Meterological Society*, **93**, 501–8.

HOLMES, R. M. [1961], Estimation of soil moisture content using evaporation data;

In Proceedings of Hydrology Symposium No. 2. Evaporation. Department of Northern Affairs and National Resources (Ottawa), pp. 184–96.

KING, K. M. [1961], Evaporation from land surfaces; In *Proceedings of Hydrology Symposium No. 2. Evaporation*, Department of Northern Affairs and National Resources (Ottawa), pp. 55–80.

MATHER, J. R. [1963], Average Climatic Water Balance Data of the Continents, No. 2 Asia (excluding U.S.S.R.); *Publications in Climatology XVI. No. 2* (Centerton, New Jersey), 262 p.

MCCULLOCH, J. S. G. [1965], Tables for the rapid computation of the Penman estimate of evaporation; *East African Agricultural and Forestry Journal*, **30**, 286–95.

MONTEITH, J. L. [1965], Evaporation and Environment; *In The State and Movement of Water in Living Organisms*, Society for Experimental Biology, 19th Symposium (Cambridge), pp. 205–34.

MORTON, F. I. [1967], Evaporation from large deep lakes; *Water Resources Research*, **3**, 181–200.

OLIVIER, H. [1961], *Irrigation and Climate* (London), 250 p.

PELTON, W. L. [1961], The use of lysimetric methods to measure evapotranspiration; In *Proceedings of Hydrology Symposium No. 2. Evaporation*, Department of Northern Affairs and National Resources (Ottawa), pp. 106–22.

PENMAN, H. L. [1963], *Vegetation and Hydrology*; Technical Communication No. 53, Commonwealth Bureau of Soils (Farnham Royal), 124 p.

PERRIN DE BRICHAMBAUT, G. and WALLÉN, C. C. [1963], *A study of Agroclimatology in Semi-Arid Zones of the Near East*; World Meteorological Organization, Technical Note No. 56 (Geneva), 64 p.

RUTTER, A. J. [1967], Evaporation in Forests; *Endeavour*, **26**, 39–43.

SELLERS, W. D. [1956], *Physical Climatology* (Chicago), 272 p.

THORNTHWAITE, C. W. [1948], An approach towards a rational classification of climate; *Geographical Review*, **38**, 55–94.

THORNTHWAITE, C. W. and HARE, F. K. [1965], The loss of water to the air; In Waggoner, P. E., Editor, *Agricultural Meteorology, Meteorological Monographs*, 6, No. 28 (Boston, Mass.), pp. 163–80.

WALLÉN, C. C. [1966], Global solar radiation and potential evapotranspiration in Sweden; *Tellus*, **18**, 786–800.

WALLÉN, C. C. [1967], Aridity definitions and their applicability; *Geografiska Annaler*, Ser. A, **49**, 367–84.

WORLD METEOROLOGICAL ORGANIZATION [1966], Measurement and estimation of evaporation and evapotransportation; *Technical Notes No.* 83.

4.II. Soil Moisture

M. A. CARSON

Department of Geography, McGill University

A large literature exists on the physics of soil moisture and upon the way in which soil water influences the nature of the soil in which it exists. This essay is not intended as a summary of these topics. Such standard works as *Soil Physics*, by Baver [1956], and *Soil*, by Jacks [1954], already serve this purpose admirably. The purpose of this essay is to outline the part played by soil moisture in some aspects of the denudation of the landscape.

1. The nature of soil moisture

A number of forces are capable of attracting water into dry soil. One is the simple affinity of the soil particles for water vapour in the soil atmosphere, although such hygroscopic water forms only a very small percentage of the water existing in most soils. A more important mechanism is the capillary suction which exists on the menisci of water films in contact with soil particles. This suction is usually explained by analogy with the rise of water in a capillary tube. Insertion of a thin

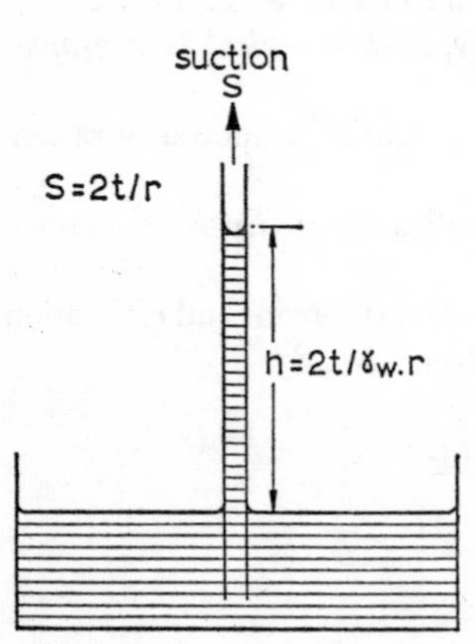

Fig. 4.II.1 Suction on water in a capillary tube.

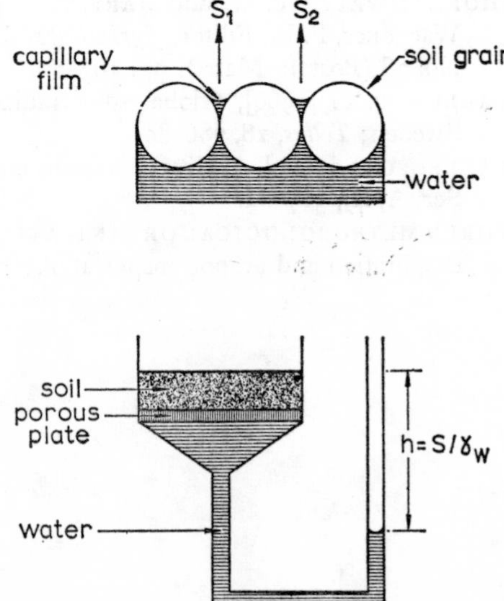

Fig. 4.II.2 Suction on water in soil.

tube into a tank of standing water (fig. 4.11.1) produces a rise in the level of the water in the tube relative to the level outside it. The extra height of the water in the tube is attributable to capillary suction acting against the force of gravity, which, alone, would maintain the same level inside and outside the tube. The magnitude of the suction is determined by the surface tension of the water and the radius of the meniscus. The height of capillary rise in the tube is determined by the ratio of the suction and the weight of a unit volume of water. The menisci in the pores of soil possess similar suction (fig. 4.11.2), and this is also measured by noting the amount of water displaced against gravity. As capillary water is

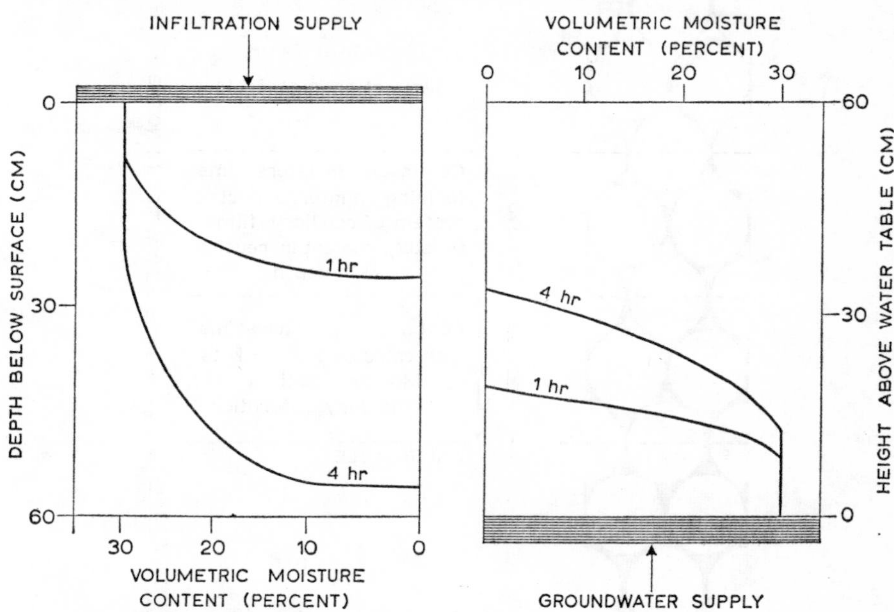

Fig. 4.11.3 Distribution of water in a sandy soil mass during (*left*) surface infiltration, and (*right*) capillary rise from a water table (After Liakopoulus, 1965a).

drawn out of a soil mass, the water which remains in the soil occupies smaller and smaller pores, and the suction on this water increases in precisely the same way as the height of rise in a capillary tube increases with decreasing radius of the tube. The capillary suction in a soil, together with the gravity force, determines the major movement and distribution of water in soil.

The pattern of change in the distribution of water in a soil mass during infiltration from water on the surface, and also during the entry of water upward from a ground-water system, has been treated by many workers. The similarity between the two processes has been emphasized by Liakopoulos [1965a] and is demonstrated in fig. 4.11.3. The entry of water into a soil mass from the surface proceeds by the downward advance of a wetting front where there is a sudden change from wet to dry soil. The amount of moisture in the soil shows little change with depth at a distance behind the wetting front, although it decreases rapidly in the area immediately behind it. A similar pattern of change occurs with the upward advance of a wetting front from a ground-water supply. In the soil just beneath the wetting line there is an increase in water content with depth until a constant moisture content is attained. This special value is identical in the two cases, and in the sandy soil used by Liakopoulos was about 30% by volume, which represents about three-quarters saturation of the pore space. These results agree with those of Bodman and Colman [1943] and other early workers.

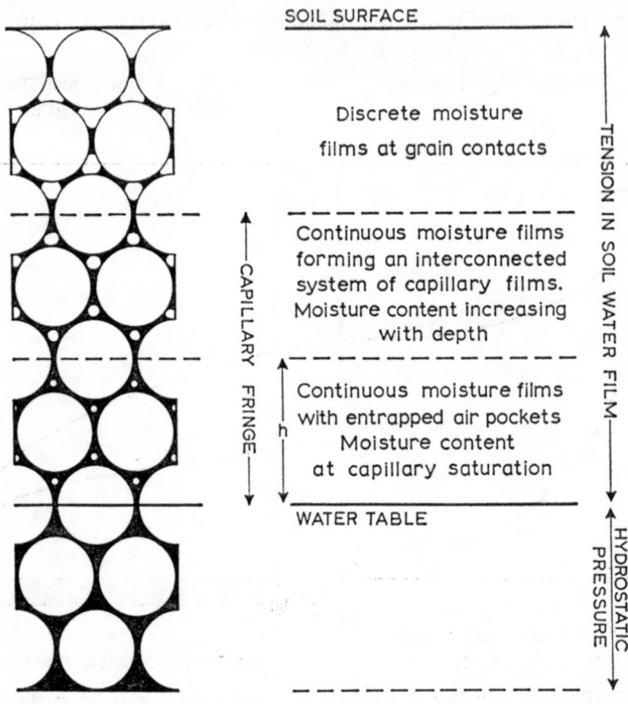

Fig. 4.II.4 The zones of capillary moisture in soil.

The ultimate equilibrium position to which the system tends differs in the two cases. In the case of capillary rise (fig. 4.II.4) a number of distinct zones exist. Immediately above the water-table is a belt of soil with constant moisture content at capillary saturation. Above this there is a systematic decrease in the amount of moisture in the soil. The area of continuous capillary water above the water-table is termed the capillary fringe, and above it the moisture films separate into discrete menisci. The thickness of the capillary fringe depends upon the size of the pores in the soil mass in much the same way as the height of water in a capillary tube depends upon the radius of the menicus. The theoretical extent of the capillary fringe is very great in the case of a clay with very small pores, whereas it is negligible in a sandy soil. The rate of capillary flow in a soil with very small pores is usually so slow, however, that the state of ultimate equilibrium is never attained, although, even then, the extent of this zone is appreciably greater than in sandy soils.

The ultimate moisture distribution in the infiltration case must differ from the capillary rise case, since the water supply will eventually exhaust itself. The pattern of moisture redistribution after the stage when all the water has entered the soil mass is unfortunately not clearly known. All the water above the wetting front will not continue to drain down through the soil. The gravity force will draw some water through the initial wetting front, but the development of menisci and capillary suction in the soil mass will tend to oppose this force. This

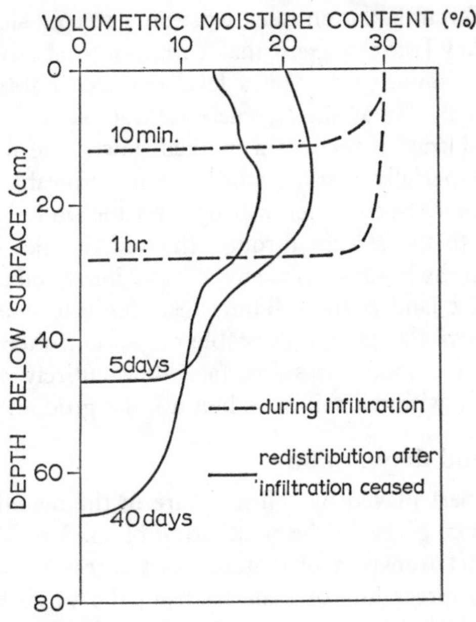

Fig. 4.11.5 Distribution of water in a sandy soil mass during and after infiltration from the surface (After Liakopoulos, 1965b).

was noted by early workers who distinguished between 'movable' and 'immovable' water. A soil which has just shed all its drainable water and still retains its maximum capacity of water which is immovable under the attraction of gravity alone was stated to exist at its *field capacity*. In this state the soil water would be held by a suction of about 20–40 in. of water. The amount of moisture in the soil at field capacity would depend upon the number of small pores and was thought to range from about 30% by volume for a clay soil to less than 10% for sandy soils. The practical value of the concept of field capacity has been criticized by a number of workers. The tests by Liakopoulos [1965b] show that even in a fine sand (fig. 4.11.5) drainage past the initial wetting front proceeds a long time after infiltration has ended.

The water which drains past the initial line of the wetting-front moves towards an underlying ground-water system. During this movement the water is still subject to lateral capillary suction and, as pointed out by Sherman [1944], it may never reach the water-table unless most of the soil pores are large. This depends also on the height of the water-table. The subsequent history of movable water beneath the initial wetting front depends, in addition, on the nature of the solid rock underlying the soil mantle. A shattered mass of rock with large gaps will facilitate downward percolation to the water-table, whereas a highly impermeable stratum may lead to a temporary ground-water system above the main one at depth.

The picture presented above is a simplified account of the movement and distribution of water in a soil mass during infiltration from the surface on level ground. A very different situation must exist on a steeply sloping land surface. The tests by Whipkey [1965] suggest that a wetting front advances into the soil mantle in much the same way as on a level surface. These tests do suggest, however, that the soil may attain complete saturation. Another feature is that during and after infiltration there is a marked downslope flow within the soil mass, and this is especially marked where an impermeable soil layer retards vertical entry of water. Such water will by-pass the underlying ground-water system and return to the stream through the soil mantle. It is important to realize therefore that the infiltration of storm water into a soil may differ radically on slopes from a flat land surface. These tests indicate that temporary water systems perched above the main water-table may occur often during prolonged rainstorms and offer a mode of subsurface flow entirely different from the pattern which occurs within a soil mass beneath flat ground.

2. Soil moisture and denudation

The most obvious part played by soil moisture in the denudation of the landscape is the assistance given in the weathering of solid rock masses into loose debris and the direct transport of material as solutes and colloids out of the waste mantle. This direct loss of material from the waste mantle by moving water has attracted little interest in the past. Attention has been focused upon the conditions which induce subsequent redeposition at another level within the mantle and thus create a soil profile.

There is never uniformity with depth in a soil mantle. The upper parts of the mantle are more humic, and the lower parts tend to be more moist. Soil minerals which are unstable in the upper levels and taken into solution by the soil water may attain greater stability at depth in the mantle and redeposition occurs. This eluviation and illuviation inevitably accentuate the differences between the upper and lower parts of the mantle, and a soil profile ensues. A well-leached soil commonly shows a pallid layer of silt and sand grains which overlies a horizon of illuviated clay and other minerals. The emergence of a soil profile is complicated in dry areas by the upward movement of soil water and the deposition of salts in the upper crust with evaporation.

The classic issue of the development of soil profiles usually assumes a flat surface. The loose material which mantles most hillslopes often shows, in contrast, little sign of illuviation: the only noticeable change with depth is an increase in the amount of unweathered debris. The absence of a soil profile on slopes may be due to many reasons. The soil mantle on a hillslope is subject to continual erosion of different types, and it is maintained only by the compensating supply of new soil through the weathering of the underlying solid rock. Although a soil mantle may exist permanently on a slope, it exists in a dynamic state (Nikiforoff, 1949), and a particular mass of soil may not stay on the slope for a sufficiently great length of time to develop a profile. Another explanation may be the tendency of soil water to seep downslope rather than vertically, as noted by Whipkey [1965], and under these circumstances a downslope change in soil type rather than a vertical profile might materialize. Such a pattern would

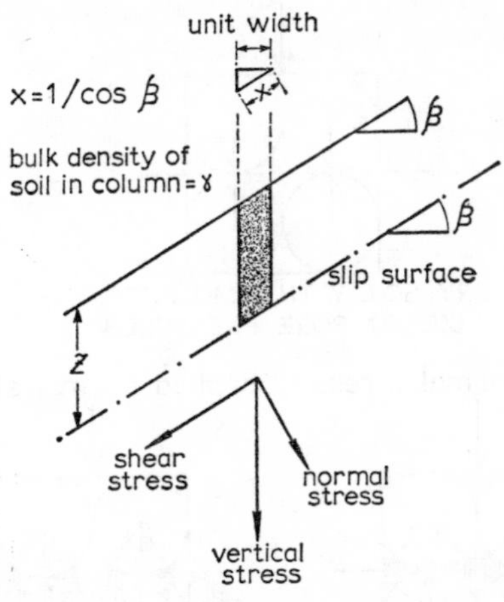

Fig. 4.II.6 The stresses imposed by gravity at a point in a soil mass.

be comparable to the catena sequence, except that it would be the product of subsurface soil water.

The transport of material out of the soil mantle by moving water within the soil mass is only a minor process, at least directly, in the denudation of the landscape. A residual soil mass will always survive this process, and much of the material lost from one part of the mantle may be expected to be deposited in another part. The major landforms are shaped by the processes of soil wash and mass-movement of different types which act upon this residual mantle. Soil moisture has a distinctive role in each of these. The influence of antecedent soil moisture in determining the amount of surface runoff and the extent of soil wash is discussed in Chapter 5.I. The less obvious role of soil moisture in the mass movements of shallow landslides and seasonal soil creep is discussed here.

Soil moisture plays a vital role in determining the stability of a soil mantle on a hillside. The shear stresses which exist within a soil mass and the shear strength to withstand these stresses both depend partly upon the amount of moisture in the soil. The shear stress on the plane of failure in fig. 4.II.6 is dependent upon the angle of the slope, the depth of the plane, and the density of the soil mass above it; the last of these will vary with the amount of moisture in the soil. The shear strength of a soil mass is derived from a cohesive element and a frictional component. The amount of internal friction developed along a plane of failure depends upon the effective stress normal to that plane as well as the angle of shearing resistance of the soil. The effective normal stress itself depends not only upon the component of the weight of the soil column at right angles to the

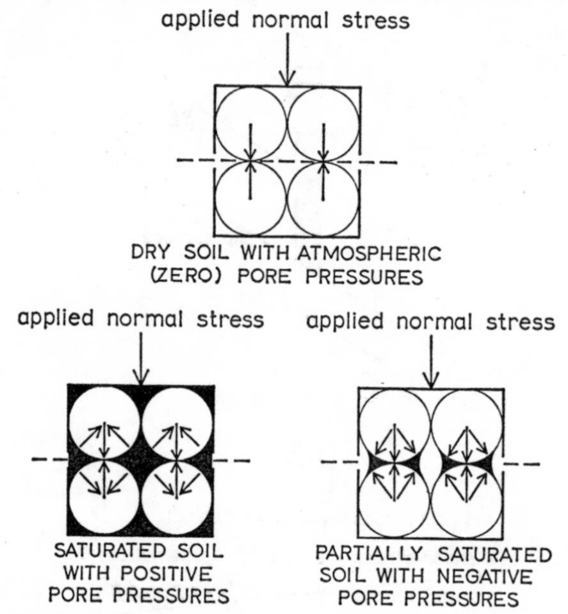

Fig. 4.II.7 The influence of moisture on pore pressures in soil.

plane (fig. 4.II.6) but also upon the pressure in the pores of the soil. This is shown in fig. 4.II.7. In a soil mass which is completely dry the normal stress applied by the overlying material is neither supplemented nor alleviated by the air pressure in the pores, since the pressure is atmospheric. When the soil pores are partly filled with water the pressure in the water films under the menisci is less than atmospheric (Skempton, 1960), so that the overall pore pressure is negative and a suction force augments the applied normal force in drawing the soil grains together. In a soil mass which is saturated with free-draining water the pore pressures are positive relative to the atmospheric datum, and this acts against the applied normal stress. The moisture content of the soil, through its influence on the effective normal stress and thus on the amount of internal friction which may be developed upon a potential plane of failure, is a vital consideration in the stability of a hillside soil mantle. It has indeed been suggested by Vargas and Pichler [1957] that the majority of natural landslides owe their origin to the development of positive pore pressures in a soil mass during prolonged rainstorms.

An implication of this is that, assuming that a soil mass is typified by particular pore-pressure values, the angle of limiting stability in any area will congregate around particular values determined by the pore pressure and the shear strength of the soil. The maximum angle of a stable slope occurs when the shear stress and the shear strength of the soil are just balanced. In the situation depicted in fig. 4.II.6 this is given by:

$$\gamma \,.\, z \,.\, \sin \beta \,.\, \cos \beta = c' + (\gamma \,.\, z \,.\, \cos^2 \beta - u) \tan \phi'$$

where γ is the density of the soil mass;
z is the depth of the plane of failure;
β is the angle of slope;
c' is the cohesion of the soil;
u is the pore pressure;
ϕ' is the angle of shearing resistance of the soil.

The effect of pore pressure is most conveniently illustrated by dealing with soils which have negligible cohesion. The maximum stable slope in the situation in fig. 4.II.5 is then given by:

$$\tan \beta = \tan \phi' \,(1 - u/(\gamma \,.\, z \,.\, \cos^2 \beta))$$

Soil mantles which never attain complete saturation and never fully dry out will always possess negative values of u, the pore pressure, and may thus stand at angles which exceed the angle of shearing resistance of the soil material. Schumm [1956] suggested that this occurs on 40–45-degree badland slopes in South Dakota, where there is sufficient silty material to provide lasting capillary suction. In contrast, soils which are essentially loose rock fragments are unlikely to attain complete saturation when they mantle hillsides due to the large pores, and for the same reason they are unlikely to maintain capillary water films permanently. They are thus characterized by pore pressures which are essentially atmospheric ($u = 0$), and the maximum stable slope is in this case the same as the angle of shearing resistance of the material. This is very probably the reason why so many scree slopes exist at angles near to 35 degrees, since this value approximates the angle of shearing resistance of small loose rocky matter. The majority of hillslopes, however, at least in humid areas, stand at angles which are less than the angle of shearing resistance of the soil mantle, and it seems very likely that this is due to the development of positive pore water pressures at times of prolonged rainstorms which give rise to perched water-tables. The pore-water pressures in free-draining water depend upon the pattern of flow, but in the case of ground-water flow parallel to the surface (fig. 4.II.8) the pressure at any point is given by:

$$u = \gamma_w \,.\, z \,.\, \cos^2 \beta$$

where γ_w is the density of water.

Substitution of this value in the previous equation gives:

$$\tan \beta = \tan \phi' (1 - \gamma_w/\gamma)$$

and since the bulk density of most surface soils is about twice the density of water, this indicates that the maximum stable slope under these circumstances should approximate to half the angle of shearing resistance in tangent form. The work of Skempton and DeLory [1957] suggests that, in the London Clay at least, this hypothesis is supported by the field evidence: angles of limiting slope approximate 8–9 degrees, and this agrees with a residual angle of shearing

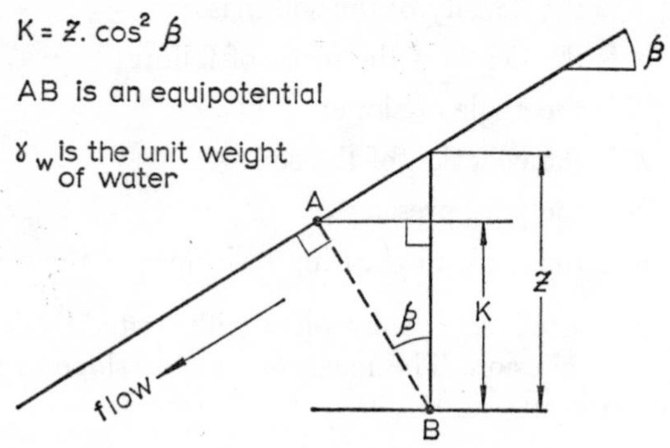

Fig. 4.II.8 The relation between pore-water pressure and depth in a soil mass with ground-water flow parallel to the surface.

resistance which is near to 16 degrees. In the case of an area which experiences widespread artesian pore-water pressures the limiting slope angle will be less than the value predicted on the basis of the previous model, although such pressures are possibly rather rare.

Soil moisture thus plays a conspicuous part in the denudation of the landscape under landslides and, through the attainment of particular values of pore pressures, may lead to the emergence of special angles of limiting slope in any one area. A hillside which is stable against rapid mass-wasting is not immune to still further transport of debris downslope, and in humid areas subsequent denudation is mostly through moisture-induced soil creep.

There are two types of soil creep which act upon hillslopes. One is the unidirectional movement downslope under the impetus directly of gravity and designated by Terzaghi [1950] as shear creep. The rate of this type of movement is probably very slow. Superimposed upon this continuous creep is seasonal soil creep. This is produced by random and cyclic disturbances which operate on flat land as well as on slopes. In the latter case, however, they are given a systematic bias downslope by the component of the gravity force. The major mechanism underlying movement in seasonal soil creep in humid temperate areas is the expansion and shrinkage of soils with changes in the moisture content of the soil. Some pioneer tests by Young [1960] indicated that under this mechanism alone soil creep might be expected to transport 0·5–1·0 cm^3 of soil past a line 1 cm wide across the slope in an average year. Subsequent work (Young, 1963) suggests that this approximates the actual amount of soil creep on slopes in the British Isles, and the conclusion that the bulk of seasonal soil creep is due to changes in the amount of moisture in the soil is supported by other workers. The creep rate clearly depends very much on the soil type on the slope. Sandy soils swell very little and clays considerably when water is absorbed, and it might be expected that the creep rate would increase with the amount of colloidal material in the soil. This is a possibility, although other features, such as angle of slope, may influence the rate of soil creep.

One of the major controversies in geomorphology is whether the processes of denudation cause hillslopes to decline in steepness or retreat at an unchanging angle. It is doubtful whether soil creep will produce either of these two changes on a straight hillside. The decline or the retreat of a straight hillslope demands that there is a systematic increase in the discharge of soil moved at different points downslope. This, in the case of soil creep, means that there must be one or both of an increase in the rate of soil creep and an increase in the thickness of the moving mantle downslope. The meagre evidence available suggests that neither of these occurs and, by implication, a straight slope acts essentially as a plane of transport for material from upslope with no effective erosion on the straight slope. The product of this is the replacement of the straight slope by the encroachment downslope of it of a convex hilltop, and there is neither retreat nor decline in the straight slope during this sequence.

The evolution of a landscape which is stable against rapid mass-wasting is, at least when it is free from the action of moving ice and the particular effects of karst limestone, determined primarily by the strength of soil wash as against soil creep. The balance between these two processes may depend a great deal on the soil type. Schumm [1956] showed this in the badlands of South Dakota and Nebraska: some clays absorbed all surface water and were subject only to creep while more impermeable clays induced surface run-off and soil wash. The effect of micro-climate on the balance between the two processes was discussed by Hack and Goodlett [1960]: moist slopes in the Appalachians appeared to be subject mostly to creep, while soil wash dominated on dry slopes. The major distinction between the two processes undoubtedly occurs with the differences in climate on the world scale. Soil creep in semi-arid areas is usually dwarfed in importance by the vast amount of soil wash which occurs in torrential storms on bare hillslopes. The results of work in humid areas suggests, in contrast, that soil creep is far more important and that it is only in the most intense and prolonged rainstorms that run-off and soil wash may take place. It is perhaps not mere coincidence, therefore, that the slopes of arid areas are dominated by straight profiles and sharp crests, while the upper hillslope convexity assumes its most distinctive development in the humid temperate areas of the earth.

The existence of soil moisture is clearly fundamental to the major processes of denudation. It is not surprising that differences in soil type and climate, affecting the amount and movement of soil moisture, are translated into differences in the efficacy of these processes and thus mirrored in the earth's landforms.

REFERENCES

BAVER, L. D. [1956], *Soil Physics*; 3rd Edn. (New York).

BODMAN, G. B. and COLMAN, E. A. [1943]. Moisture and energy conditions during downward entry of water into soils; *Proceedings of the Soil Science Society of America*, **8**, 116–22.

HACK, J. T. and GOODLETT, J. C. [1960], Geomorphology and forest ecology of a mountain region in the central Appalachians; *U.S. Geological Survey Professional Paper* 347.

JACKS, G. V. [1954], *Soil* (Edinburgh), 221 p.

LIAKOPOULOS, A. C. [1965a], Theoretical solution of the unsteady unsaturated flow problem in soils; *Bulletin of the International Association of Scientific Hydrology*, **10**, 5–39.

LIAKOPOULOS, A. C. [1965b], Retention and distribution of moisture in soils after infiltration has ceased; *Bulletin of the International Association of Scientific Hydrology*, **10**, 58–69.

NIKIFOROFF, C. C. [1949], Weathering and soil evolution; *Soil Science*, **67**, 219–30.

SCHUMM, S. A. [1956], The role of creep and rainwash on the retreat of badland slopes; *American Journal of Science*, **254**, 693–706.

SHERMAN, L. K. [1944], Infiltration and the physics of soil moisture; *Trans. American Geophysical Union*, **25**, 57–71.

SKEMPTON, A. W. [1960], Effective stress in soils, concrete and rocks; *Pore Pressure and Suction in Soils*, 4–16.

SKEMPTON, A. W. and DELORY, F. A. [1957] Stability of natural slopes in London Clay; *Proceedings of the 4th International Conference of Soil Mechanics*, **2**, 378–81.

TERZAGHI, K. [1950], Mechanism of landslides; *Bulletin of the Geological Society of America, Berkey volume*, 83–122.

WHIPKEY, R. Z. [1965], Subsurface storm flow from forested slopes; *Bulletin of the International Association of Scientific Hydrology*, **10**, 74–85.

VARGAS, M. and PICHLER, E. [1957], Residual soil and rock slides in Santos (Brazil); *Proceedings of the 4th International Conference of Soil Mechanics*, **2**, 394–8.

YOUNG, A. [1960], Soil movement by denudational processes on slopes; *Nature*, **188**, 120–2.

YOUNG, A. [1963], Soil movement on slopes; *Nature*, **200**, 129–30.

5.I. Infiltration, Throughflow, and Overland Flow

M. J. KIRKBY
Department of Geography, Bristol University

When the rainfall that has not been intercepted by vegetation reaches the ground surface part of it fills small surface depressions (depression storage), part percolates into the soil, and the remainder, if any, flows over the surface as overland flow. Each component of this equation is highly variable, and depends not only on the intensity of the rainfall but also on soil, vegetation, and surface gradient. The amount of water intercepted by vegetation depends on the type of plants and their stage of growth, but it is usually close to a value given by all of the first 1·0 mm of rainfall and 20% of the subsequent rainfall in any one storm. Rainfall reaching the soil surface has to fill the small depressions on the surface before any overland flow can occur, even on a totally impermeable surface. This depression storage does not vary with the amount of rainfall but with the nature of the surface, especially with slope gradient, vegetation cover, and land-use practices. Under natural conditions depression storage absorbs about 2–5 mm of rainfall in any one storm. Contour ploughing is particularly effective in increasing depression storage by as much as ten times.

1. Infiltration

Infiltration rate is defined as the maximum rate at which water can penetrate into the soil. For initially moister soils the infiltration rate is lower throughout storms, and for all soils it decreases during the course of a storm. The rate at which water can travel through the soil depends on the number and size of pore spaces in the soil and the distribution of water within them. In effect, the infiltrating water has two components, a transmission component, which is constant and represents a steady flow through the soil; and a diffusion component, which is an initially rapid, and then an increasingly slow, filling-up of air-filled pore spaces, from the surface downwards. These components can be expressed in the infiltration equation (Philip, 1957):

$$f = A + B \, . \, t^{-\frac{1}{2}} \qquad (1)$$

where f is the instantaneous rate of infiltration;
t is time elapsed since the beginning of rainfall;
A is the 'transmission constant' of the soil; and
B is the 'diffusion constant' of the soil.

In this equation the transmission and diffusion terms can be identified with the two components of an idealized model. The transmission term represents unimpeded laminar flow through a continuous network of large pores. The diffusion term represents flow in very small discrete steps from one small pore space

TABLE 5.1.1 Variation of minimum infiltration rates with soil grain size, initial moisture content, and vegetation cover

(*a*) The effect of grain size in initially wet soils without vegetation cover

Grain size class	*Infiltration rates* (mm/hr)
Clays	0–4
Silts	2–8
Sands	3–12

(*b*) The influence of moisture content for Illinois clay-pan soils (after Musgrave and Holtan, 1964)

Initial moisture content (%)	*Infiltration rates* (mm/hr)		
	Good grass cover		*Poor weed cover*
	Topsoil > 13 in. thick	*Topsoil < 13 in. thick*	*Topsoil < 13 in. thick*
0–14	17	19	6
14–24	7	7	4
24 +	4	4	3

(*c*) The influence of ground cover for Cecil, Madison, and Durham soils (after Musgrave and Holtan, 1964)

Ground cover	*Infiltration rate* (mm/hr)
Old permanent pasture	57
Permanent pasture; moderately grazed	19
Permanent pasture; heavily grazed	13
Strip-cropped	10
Weeds or grain	9
Clean tilled	7
Bare ground crusted	6

to the next, in a random fashion. The only reason that a net diffusion flow results is that more pores are dry lower down, so that there is greater opportunity for downward movement than for upward. In an actual soil the two phases of this model cannot be separated, as all pores show a combination of the two types of behaviour, but equation (1) remains a good approximation for measured infiltration rates.

Table 5.1.1 shows some of the range of variation which can be expected in infiltration rates under a range of vegetation and moisture conditions, and in fig. 5.1.1 a comparison is made between typical infiltration rates and expected storm rainfall rates.

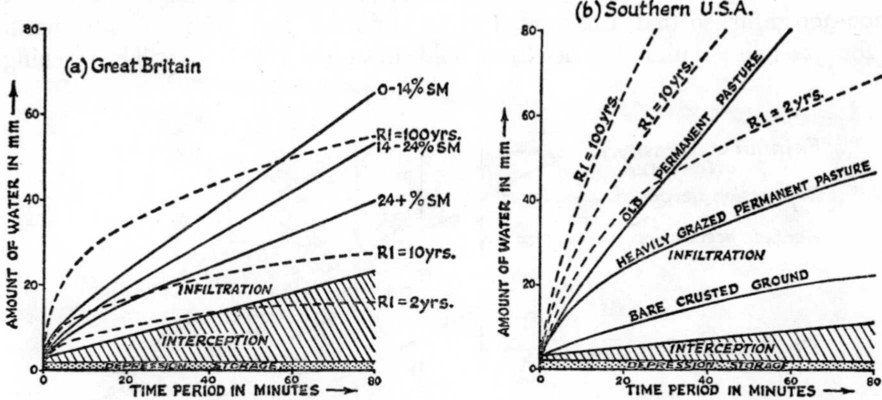

Fig. 5.1.1 Comparison of expected rainfall intensities with infiltration rate, interception, and depression storage, with representative values to demonstrate the relative frequency of overland flow in Great Britain and southern United States.

(*a*) Great Britain: rainfall (Data from Bilham, 1936), interception values for 100% vegetation cover, infiltration rates for Illinois clay-pan soils with good grass cover (Data from Musgrave and Holtan, 1964).

(*b*) Southern United States: rainfall (Data from Yarnell, 1935), interception values for 50% vegetation cover, infiltration rates for Cecil, Madison, and Durham soils (Data from Musgrave and Holtan, 1964).

2. Overland flow and throughflow

'Horton overland flow' is defined as overland flow which occurs when rainfall intensity is so great that not all the water can infiltrate, and is described by Horton [1945]. This type of overland flow is a fairly common phenomenon in semi-arid climatic conditions, but is relatively rare in humid and humid-temperate conditions. The role of vegetation is thought to be a critical cause of this distinction. Vegetation increases the infiltration rate by promoting a thicker soil cover, a better soil texture, and by breaking the impact of raindrops on the surface. Its effect on soil structure is mainly to build up an organic-rich A horizon with a relatively open pore structure and high permeability. If raindrops strike the surface without being impeded by vegetation fine material is thrown into suspension by the impact and is redeposited as an almost impermeable surface skin which can lower infiltration by as much as ten times. Vegetation therefore has a controlling influence on Horton overland flow by increasing both the initial depression storage and the infiltration rate, so that where a dense vegetation cover is established Horton overland flow is very unusual. Soil compaction by animals and vehicles reduces the infiltration rate while increasing depression storage, so that its net influence is problematic.

Within a small drainage basin, where soils are more or less homogeneous, it may be expected that interception, depression storage, and infiltration rate will not vary greatly, so that the conditions for Horton overland flow will be satisfied by comparable intensities and durations of rainfall, and overland flow will occur simultaneously all over the basin. Typical velocities for overland flow are about 200–300 m/hr, so that in a rainfall of 1 hour water from all points of a basin (200–300 m is a typical distance from divide to stream in Britain) will be reaching

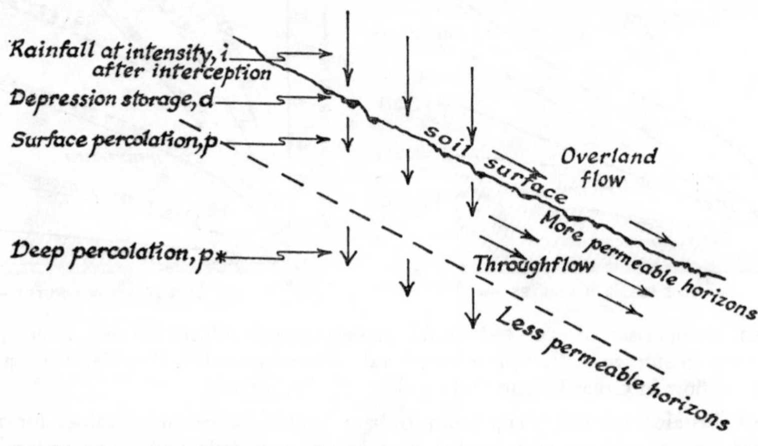

Fig. 5.1.2 Flow components over and within the soil.

a stream channel, and flow over the slope surface will have reached a steady state, represented by the equation:

$$q_0 = (i - f) \,.\, a \tag{2}$$

where q_0 is the overland flow discharge per unit contour length;

i is the rainfall intensity after interception;

f is the infiltration rate; and

a is the area drained per unit contour length (equal to the distance from the divide if all the contours are straight lines).

Thus, provided that the rainfall intensity is high enough (or the infiltration rate low enough) for this type of overland flow the actual magnitude of the flow will be strongly dependent on the area or distance of overland flow, and will be almost independent of the storm duration, provided that it exceeds a reasonable minimum value. This is Horton's [1945] classic overland flow model.

Some of the water which infiltrates into the soil passes downward to recharge the water-table, and some, usually the greater part, flows down the hillside within the soil layers as 'throughflow' and ultimately contributes to streamflow (fig. 5.1.2). Within the soil, permeability varies and is generally highest in the open-textured organic A_0 horizon and the eluviated A_1 horizon. B horizons tend to be less permeable because clays are washed down into them, and in some cases because of the development of a hardpan. Conditions in the parent bedrock vary widely from limestones, with open solution fissures, to totally impermeable consolidated clays and shales. Wherever permeability is decreasing downward within the soil, and this occurs most commonly at the base of the A horizon, part of the water which is percolating downwards cannot penetrate into the lower layers fast enough, and is deflected laterally within the upper layer, as throughflow. This is similar to the production of Horton overland flow at the surface, except that within the soil the reduction of permeability is usually gradual, leading to a progressive deflection of throughflow. Table 5.1.2 shows an example of the progressive decline of permeability through the soil profile for a hillslope in Ohio, U.S.A.

TABLE 5.1.2 Variations of permeability and soil type with depth in an Ohio forest soil at a gradient of 15° (after Whipkey, 1965)

Soil depth (cm)	*Textural class*	*Bulk density* (gm/cc)	*Saturated permeability* (mm/hr)
0–56	Sandy loam	1·33	—
56–90	Sandy loam	1·41	286
90–120	Loam	1·78	17
120–150	Clay loam	1·80	2

If rainfall continues for a long time soil layers become saturated and throughflow is deflected closer and closer to the surface, so that the upper, more permeable soil layers are filling up from their bases because the throughflow is unable to carry away the water fast enough. In time the soil will become saturated right up to the surface and 'saturation overland flow' will occur. Under steady rainfall this condition will ultimately be attained under rainfall intensities much lower than are required to produce Horton overland flow. Since soil thickness and the velocity of throughflow vary much more over a small area than does permeability, and since the base of a slope tends to become saturated sooner than the divide, certain parts of a hillside are likely to produce saturation overland flow preferentially, in contrast to the rather widespread production of Horton overland flow, when it occurs at all.

Throughflow, travelling through soil pore spaces rather than over the ground surface, moves at very much lower velocities than overland flow. Rates of 20–30 cm/hr for throughflow are of the order of a thousand times lower than overland flow rates, so that periods of about 1,000 hours rainfall are needed for a steady state of flow to be achieved throughout an average basin. In practice, such a steady state is never attained for throughflow, and equation (2) for overland flow must be replaced with equation (3) for throughflow:

$$q_T = (p - f_*) \, . \, v \, . \, t \qquad (3)$$

where q_T is the throughflow discharge per unit contour length;
p is the rate of surface percolation, equal to i or f, whichever is smaller;
f_* is the rate of infiltration at the base of the more permeable soil;
v is the velocity of throughflow; and
t is the time elapsed (strictly $v \, . \, t$ should be replaced by the area/unit contour length within a distance $v \, . \, t$ upslope).

In this equation the *time elapsed* is the most important control over the flow, in place of the *distance from the divide* for Horton overland flow. In reality, the flows are delayed by the time of transmission from the surface to a zone of decreasing permeability, a period which may be a matter of minutes or a few hours.

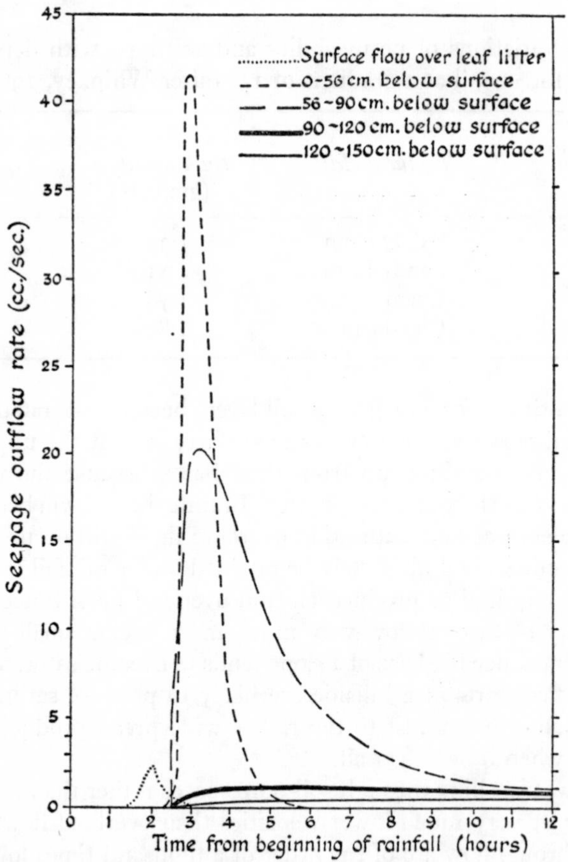

Fig. 5.1.3 Discharge hydrographs of flow within the soil resulting from a simulated storm of 5·1 cm/hr, lasting 2 hr, on a 16° slope which had previously drained for more than four days. The rapid, although small, Horton overland flow results from the initially low permeability of the dry surface soil, which rapidly increases with wetting. The lag before throughflow begins is the time taken for rain to infiltrate vertically to the 90-cm-deep, less-permeable interface (After Whipkey, 1965, p. 81).

If outflow from various depths in the soil is measured throughout a storm of uniform intensity there will be an initial transmission period of little or no flow, followed by a period in which the flow is increasing rapidly (though not linearly, because v in equation (3) is itself varying) until the moment when rainfall ceases, after which the flow will decrease more slowly than it increased. In other words, the flow should broadly resemble a flood hydrograph of a stream, despite the absence of surface run-off. In fig. 5.1.3 (Whipkey, 1965) actual values are shown for a rainfall intensity of 5·1 cm/hr falling for 2 hours on the soil whose properties are described in Table 5.1.2. It is apparent from fig. 5.1.3 that even at this high intensity Horton overland flow is negligible, and that throughflow from shallower soil layers is later than from deeper layers due to the time lag for transmission of water down through the soil, followed by saturation from the base up.

Equation (3) contains an unknown quantity, namely the velocity of throughflow, v. In order to evaluate the throughflow more exactly, Darcy's law, which states that flow through a permeable medium is proportional to the pressure gradient, must be combined with the continuity equation, which states that differences between inflow and outflow must be accommodated by changes in moisture content. For a soil layer of uniform permeability and moisture content (in depth) Darcy's law for soil on a slope is

$$Q = z \,.\, \cos\alpha \left\{ K \,.\, m \,.\, \sin\alpha - D \,.\, \frac{\partial m}{\partial x} \right\} \tag{4}$$

and the continuity equation is

$$\frac{\partial m}{\partial t} + \frac{\partial Q}{\partial x} = i \tag{5}$$

where Q is the downslope discharge measured in a horizontal direction;
x is the distance downslope measured in a horizontal direction;
K is the soil permeability (a function of soil moisture);
D is the soil diffusivity (a function of soil moisture);
α is the surface slope angle;
m is the soil moisture content;
z is the thickness of the soil layer;
i is the rainfall intensity after interception; and
t is the time elapsed.

Solution of these equations is necessarily numerical, but in a simple actual example of a uniform soil on a uniform gradient (Hewlett and Hibbert, 1966) estimates of permeability and diffusivity were as shown in Table 5.1.3; and calculated soil-moisture patterns during (*a*) uniform rainfall and (*b*) uniform drainage were as shown in fig. 5.1.4. It can be deduced from the data of Table 5.1.3 that the permeability (K) term in equation (4) is the more important for intermediate values of moisture, while the diffusivity (D) term becomes dominant at extreme low and high values of moisture. The low-moisture case is of little interest, but the high-moisture case, when the moisture content is approaching

TABLE 5.1.3 Values of permeability (K) and diffusivity (D) for varying soil moisture, computed from the results of an experiment carried out during soil drainage of a trough of soil inclined at 22° (Hewlett and Hibbert, 1963)

Moisture content (expressed as % of saturated moisture content)	*Permeability* (K) (cm/hr)	*Diffusivity* (D) (cm/hr)
68	0·000	0·0
73	0·004	3·2
78	0·081	12·5
83	0·46	19·3
88	1·58	40·8
93	4·00	100·4
98	8·67	304·0
99	10·64	517·0
100 (saturated)	12·08	Infinity

saturation, is of great hydrologic significance. In this important case soil moisture is constant at saturation, so that equation (5) becomes extremely simple, namely

$$\frac{\partial Q}{\partial x} = i \tag{6}$$

What this means in practice is that the near-saturated soil zones respond very rapidly to changes in rainfall intensity, even before overland flow begins. Most

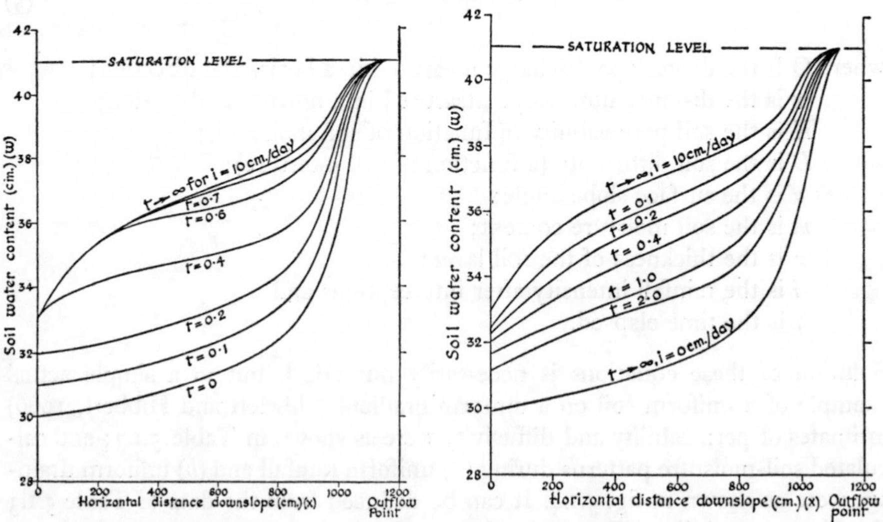

Fig. 5.1.4 Calculated moisture distributions in a 12-m-long soil trough, inclined at 22° during: (*left*) rainfall at a constant rate of 100 mm./day, and (*right*) drainage after indefinite rainfall of this intensity.

The data are calculated from the results of an experiment by Hewlett and Hibbert (1963).

important is the saturated zone which is often at the side of flowing streams, for the outflow from this zone is given by:

$$q_T = i \cdot x_s + q_T^* \tag{7}$$

where x_S is the width of the saturated zone;
q_T is the throughflow discharge per unit contour length;
q_T^* is the throughflow contribution from farther upslope; and
i is the rainfall after interception.

Equations (4) and (5) also provide a basis for assessing the influence of changing gradient or (with slight modification) of contour curvature on the throughflow discharge within the soil. The following four regions of a slope are the most likely to become saturated, and hence provide more rapid response to rainfall and more frequent saturation overland flow:

1. areas adjacent to perennial streams;
2. areas of concave upwards slope profile;
3. hollows (areas of concave outward contours);
4. areas with thin or impermeable soils.

3. Areas contributing to stream flow

The relative infrequency of overland flow in humid regions, together with the very low velocities of throughflow, suggest that most of the rainfall which falls on hillslopes is unable to reach a channel until long after the rainfall has stopped and the stream flood peak has passed. In other words, only water from a relatively small 'contributing area' is able to reach a channel in time to contribute to the flood hydrograph of the stream. In its simplest form, for a rainstorm of constant intensity, the contributing area, A_c, is defined as

$$A_c = \frac{\text{Stream discharge}}{\text{Rainfall intensity}} \tag{8}$$

Where the rainfall intensity varies during the storm its value in the equation is necessarily an average one, weighted towards the most recent intensities. The contributing area is continuously changing during the storm (fig. 5.1.5), and is generally at its greatest at about the same time as the peak discharge in the stream. For basins measured in North Carolina the maximum contributing area varied relatively little from storm to storm, but from basin to basin it ranged from 5 to 85% of the total drainage area. These measurements of contributing area can be compared with models derived from the two types of hillside flow: 1. 100% overland flow, and 2. 100% throughflow. During 100% overland flow water from all parts of the basin commonly reaches a channel within about an hour, so that for a storm lasting longer than an hour the contributing area, expressed as a percentage of the total drainage area, is

$$\frac{\text{Rainfall intensity} - (\text{Infiltration and surface losses})}{\text{Rainfall intensity}} \times 100\%$$

Under conditions where vegetation and soil are thin or absent, this contributing area may be large, and the value of 85% contributing area refers to an abandoned copper strip-mining area with less than 36% vegetation cover.

During 100% throughflow the contributing area consists of:

1. the area of the stream channels themselves, which is usually 1–5% of the drainage area;
2. the areas of saturated or near-saturated soil, mainly adjoining channels, which respond rapidly to changes in rainfall intensity as is shown in equations (6) and (7);
3. a narrow strip of hillside around the saturated areas, the width of which is determined by the slow rates of throughflow in unsaturated soils.

In the North Carolina basin with a 5% contributing area, the actual channel area and a swampy area backed up behind the stream-gauging installation

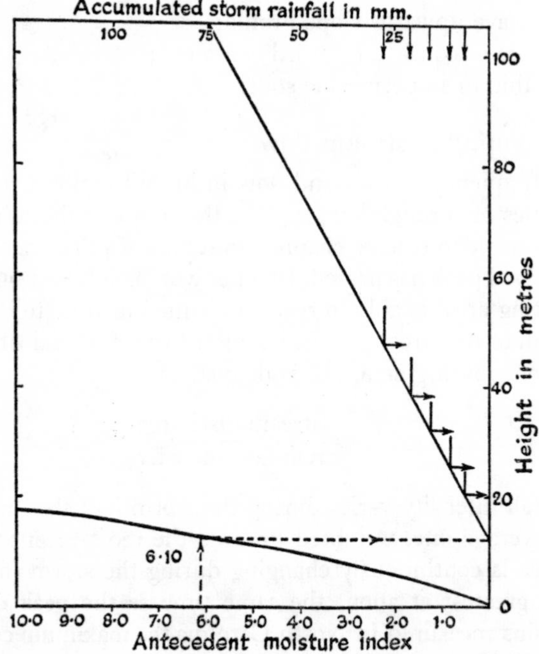

Fig. 5.1.5 An early concept of the variation of contributing area (considered as area below elevations shown), with initial moisture conditions and accumulated storm rainfall, for Bradshaw Creek, Tennessee (From T.V.A., 1964, Fig. 12).

accounted for almost the whole of the contributing area. Actual values of contributing area for humid drainage basins usually lie between these extremes, commonly in the 10–30% range, depending on soils, hillslope gradients, land use, and drainage texture.

Soils influence contributing area through their infiltration rate and the thickness of their permeable horizons, and attempts have been made to prove that contributing areas coincide with thin soil areas in a small drainage basin. Slope gradients influence the rate of throughflow (equation (4)), and hence the distribution of saturated soil and saturation overland flow. Vegetation cover and cultivation practices strongly affect the permeability of surface soil layers, mainly through the effect of vegetation in reducing rainsplash impact and through the effect of cultivation on depression storage and soil structure. In a basin with a high drainage density a relatively large area of hillslope is close to a channel, so that under throughflow the contributing area will also be relatively large; and under Horton overland flow there will be a relatively short time lag between the start of rainfall and a condition of maximum contributing area. Clearly these factors are not independent of one another, but each has a separate influence on contributing area.

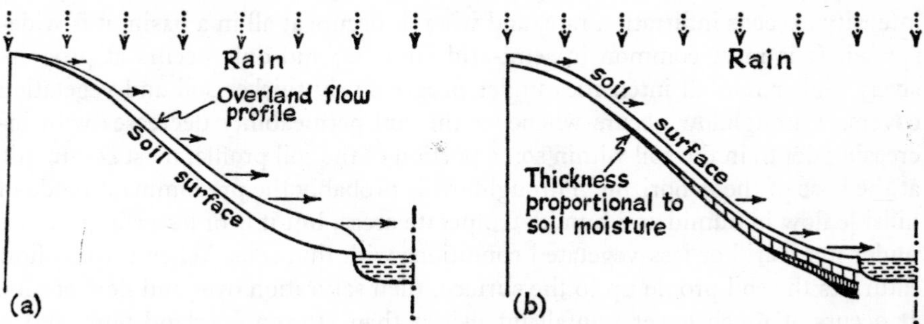

Fig. 5.1.6 Patterns of hillslope flow during Horton overland flow and throughflow. Arrow lengths show relative discharges over or through the soil.

(*a*). Horton overland flow (After Horton, 1945, p. 316). Thickness of water layer on surface is drawn proportional to actual thickness.

(*b*). Throughflow. Thickness of water layer below surface is drawn proportional to soil moisture content. Soil moisture from progressively earlier rainfalls is shown by progressively darker shading. The subsurface layer does not indicate the depth of infiltration into the soil.

Part of stream baseflow is derived from the water-table, which is itself supplied by deep percolation of water from the soil, but this contribution to baseflow is probably large only where well-defined aquifers are present. A large part of baseflow also comes from throughflow in the soil, which will take months to reach a channel from interfluve areas, and produces sufficient water to supply the measured baseflows in many areas. Since the same rain-water, much of it via throughflow, is responsible for both high and low flows in streams, all of the factors described above, which tend to produce high contributing areas during rainstorms, and hence a large proportion of total runoff during storms, also lead to reduced storage of rain-water after the flood flows have subsided, and hence to lower baseflows.

A final important contrast between the overland flow and throughflow models is that, whereas in overland flow it is the rain-water which is actually falling that flows into the stream during a rainstorm (fig. 5.1.6(*a*)), in throughflow, much of the water flowing into the channels is not physically the same as the rain-water which is currently falling (due to the time lag involved). It has been shown in infiltration experiments that almost all water flowing through the soil flows out in the order in which it flows in. This means that infiltrating water has to displace all of the soil water downslope before it can itself flow into the stream (fig. 5.1.6(*b*)), so that most water flowing into the stream, even at high flows, has been stored in the soil for a matter of weeks or months, and so has been able to come to chemical equilibrium with the soil. This soil water storage has obvious implications for interpreting the dissolved load of streams.

4. Summary

There are two extreme models of hillside water flow; the Horton overland flow model and the throughflow model. Horton overland flow occurs when rainfall intensity exceeds infiltration rate, and when it occurs at all in a basin, it is widespread. It is most common in semi-arid climates, and only occurs at progressively higher rainfall intensities under progressively thicker soil and vegetation covers. Throughflow occurs whenever the soil permeability decreases with increasing depth in the soil within some portion of the soil profile, most commonly at the base of the A horizon. Throughflow is probably the predominant mode of hillside flow in humid and humid–temperate areas, but it is of lesser importance under more arid or less-vegetated conditions with thin soils. When throughflow saturates the soil profile up to the surface, then saturation overland flow occurs. It occurs at much lower rainfall intensities than Horton overland flow, and is usually much more localized in its distribution, being commonest near streams. Under suitable conditions, both overland flow and throughflow may occur at any point although their relative frequencies will vary greatly from point to point. The separation of hillside flow into its two components, overland flow and throughflow, and a recognition of the distinct properties of each, allows a clearer understanding of the mechanisms of both streamflow and hillside erosion.

REFERENCES

BETSON, R. P. [1964], What is watershed runoff?; *Journal of Geophysical Research*, **69**, 1541–52.

BILHAM, E. G. [1936], Classification of heavy falls in short periods; *British Rainfall*, **75**, 262.

HEWLETT, J. D. and HIBBERT, A. R. [1963], Moisture and energy conditions within a sloping mass during drainage; *Journal of Geophysical Research*, **68**, 1081–7.

HORTON, R. E. [1945], Erosional development of streams and their drainage basins: hydrological approach to quantitative morphology; *Bulletin of the Geological Society of America*, **56**, 275–370.

HUDSON, N. W. and JACKSON, D. C. [1959], Results achieved in the measurement of erosion and runoff in Southern Rhodesia; *3rd Inter-African Soil Conference, Dalaba*, Paper No. 63.

JENS, S. W. and MCPHERSON, M. B. [1964], Hydrology of Urban Areas; In Chow, V. T., Editor, *Handbook of Applied Hydrology* (New York), Section 20, 45 p.

KIRKBY, M. J. and CHORLEY, R. J. [1967], Throughflow, overland flow and erosion; *Bulletin of the International Association of Scientific Hydrology*, **12**, 5–21.

LINSLEY, R. K., KOHLER, M. A., and PAULHUS, J. L. H. [1949], *Applied Hydrology* (New York), 689 p.

MUSGRAVE, G. W. and HOLTAN, H. N. [1964], Infiltration; In Chow, V. T., Editor, *Handbook of Applied Hydrology* (New York), Section 12, 30 p.

PHILIP, J. R. [1957–8], The theory of Infiltration; *Soil Science*, **83**, 345–57 and 435–48; **84**, 163–77, 257–64, and 329–39; **85**, 278–86 and 333–7.

TENNESSEE VALLEY AUTHORITY [1964], Bradshaw Creek – Elk River: A pilot study in area-stream factor correlation; *Office of Tributary Area Development, Knoxville, Tennessee*, Research Paper No. 4, 64 p. and 6 appendices.

TENNESSEE VALLEY AUTHORITY [1966], Cooperative Research Project in North Carolina: Annual Report for Water Year 1964–1965; *Division of Water Control Planning, Hydraulic Data Branch*, Project Authorisation No. 445.1, 31 p.

WHIPKEY, R. Z. [1965], Subsurface stormflow from forested slopes; *Bulletin of the International Association of Scientific Hydrology*, **10**, 74–85.

YARNELL, D. L. [1935], Rainfall intensity-frequency data; *U.S. Department of Agriculture, Misc. Publ. No. 204.*

6.I. Ground Water

J. P. WALTZ
Department of Geology, Colorado State University

1. Definition of ground water

Ground water is water which occurs beneath the surface of the earth within saturated zones where the hydrostatic pressure is equal to or greater than atmospheric pressure. This precise definition is useful in distinguishing between ground water and other types of subsurface water, such as capillary water or soil water. However, more important than fine distinctions between the various modes of subsurface water occurrence is the recognition that all water, whether in the atmosphere, on the surface, or beneath the surface of the earth, is part of a common supply. Man, with increasingly effective means to control precipitation, to create surface water storage, and to utilize the natural ground-water reservoirs for storage and development, is rapidly approaching the point where he merely has to decide where and when to establish water supplies.

Ground water is an intriguing part of the hydrologic cycle. Man cannot see water move through the ground. He can dig a hole and peer into it, but he has disturbed or destroyed part of what he wishes to see. Sophisticated electronic gadgetry may be used to 'look' beneath the surface of the earth, but the information gained by indirect measurements is often inexact and always incomplete. Hence, the movement and occurrence of ground water remain somewhat of a mystery to man.

2. Occurrence of ground water

Generally speaking, ground water can be found by drilling at almost any point on the surface of the earth *if* the hole is drilled deep enough. However, the mere presence of water is not usually what man wishes to determine. More important than the presence of ground water are the volume or *supply* of ground water and the *rate* at which the supply can be removed from the ground. Thus, our discussion of ground-water occurrence will focus first on those physical properties of earth materials which affect the amount of water which can be stored beneath the ground and the ease with which the water can be extracted from the ground.

A. Porosity and permeability

The volume of water which can be held within earth materials is controlled by the *porosity* of the materials. Porosity may be defined as follows:

$$\text{Porosity} = \frac{\text{Volume of voids in a material}}{\text{Bulk volume of the material}}$$

Porosity is usually expressed as a decimal fraction or as a percentage. For example, a rock specimen which contains pores or open spaces equal to one-fourth the total volume of the specimen would have a porosity of 25%. Naturally occurring geologic materials vary widely in porosity. Table 6.1.1 contains a list of representative porosities for various geologic materials. In the case of granular sediments, porosity is not directly affected by the *size* of the grains, but is affected by the uniformity of size, the shape, and the packing characteristics of the grains.

TABLE 6.1.1. List of representative porosities and permeabilities for geologic materials

Geologic material	*Representative porosities* (*% void space*)	*Approximate range in permeability* (gallons/day/ft²; hydraulic gradient = 1)
Unconsolidated		
Clay	50–60	0·00001–0·001
Silt and glacial till	20–40	0·001–10
Alluvial sands	30–40	10–10,000
Alluvial gravels	25–35	10,000–1,000,000
Indurated		
Sedimentary:		
Shale	5–15	0·0000001–0·0001
Siltstone	5–20	0·00001–0·100
Sandstone	5–25	0·001–100
Conglomerate	5–25	0·001–100
Limestone	0·1–10	0·0001–10
Igneous and metamorphic:		
Volcanic (basalt)	0·001–50	0·0001–1
Granite (weathered)	0·001–10	0·00001–0·01
Granite (fresh)	0·0001–1	0·0000001–0·00001
Slate	0·001–1	0·0000001–0·0001
Schist	0·001–1	0·00001–0·001
Gneiss	0·0001–1	0·0000001–0·0001
Tuff	10–80	0·00001–1

The ease with which water can move through earth materials is a function of the *permeability* of the materials. A more exact definition of permeability is given later in this chapter under the topic 'Movement of ground water'. The permeability of granular earth materials is greatly affected by the size of grains as well as by the shape, packing, and uniformity of size of grains. Permeability can be expressed in a number of different ways. For the purpose of this discussion, permeability will be described as a rate of discharge per unit area (e.g. gallons/day/ft^2) under controlled hydraulic conditions. Table 6.1.1. gives approximate ranges of permeability for various geologic materials. It is important to recognize the magnitude of the range of permeabilities for naturally occurring earth materials.

B. Aquifers

A geologic material which yields significant amounts of water to wells is called an *aquifer*. From a practical point of view, an aquifer must yield sufficient quantities of water to make it economically feasible to extract water from it. Obviously, if economics enters into the definition of an aquifer, then the intended use for the water will somewhat determine whether a water-bearing formation can be called an aquifer or not. Hence, a fractured granite may yield enough water for household uses, and would qualify as an aquifer. The same rock would probably not supply sufficient water for agricultural purposes, and in this case would not be called an aquifer.

C. The geologic framework: categories of earth materials

The study of ground water, whether for the purpose of determining its value to man or to see how it interacts with the earth through which it passes, must begin with a study of the geologic framework. Earth materials are usually classified geologically as to origin, i.e., igneous, metamorphic, or sedimentary. This classification, however, is not suited for studies of ground water. In terms of their effect on the occurrence of ground water, two broad categories of earth material are more suitable: a category which includes all unconsolidated materials, and a category which includes all indurated materials.

1. Unconsolidated materials

By definition, unconsolidated deposits do not contain cementing materials in their pore spaces. Thus, these deposits are characterized by relatively high porosities (Table 6.1.1). Included in this category of earth materials are the geologically recent deposits of alluvial or stream-transported sediments, aeolian or wind-transported particles, colluvial or gravity-driven debris, and glacial or ice-transported materials.

An analysis of recent geologic history and the origin of the various types of unconsolidated deposits in a region can be a major contribution to an evaluation of the ground-water resources. The significance of geologic studies in ground-water evaluations lies in the fact that there are predictable relations between geologic processes and the physical properties of the sediments they produce. For example, aeolian deposits are characterized by uniformity of grains in the silt and sand size range. Glacial moraines, in contrast, are poorly sorted and may contain mixtures of particles ranging from clay size to boulder size. Also, knowledge of how a material was deposited can be used to predict the overall geometry of the deposit. An example of this can be seen in the case of wind-blown sand. Because of the formation of dunes, aeolian deposits are extremely variable in thickness. Thus, if two wells, spaced 150 ft apart, were drilled in an area known to contain a buried aeolian deposit of saturated sand the first hole might penetrate 50 ft of the water-bearing sand, while the second hole might miss the deposit entirely.

2. Indurated earth materials

Indurated earth materials (rocks) also play a significant role in the occurrence and movement of ground water. Sedimentary deposits usually become indurated through cementation of the grains by a chemical precipitate of iron oxides, calcium carbonate, or some form of silica. Clay particles which might be present in a deposit also can form a cementing matrix when the deposit begins to consolidate. These cementing agents may fill essentially all of the original void spaces within the deposit. Thus, sedimentary rocks are usually much less permeable and porous than their unconsolidated counterparts. Limestones, however, provide important exceptions to this rule. Fractures in limestone may widen due to solution by ground water, making the rock highly permeable.

Igneous and metamorphic rocks are generally characterized by low permeabilities and extremely small porosities (Table 6.1.1). In most igneous and metamorphic rocks fractures are the primary source of void spaces which can contain and transmit water. Some types of volcanic rocks, however, develop high porosity due to entrapment of gas bubbles within the rock during cooling. Also, lava flows may have relatively high permeabilities because of extensive fracture systems developed during movement and cooling of the lava flow. Examples of fractured volcanic rocks which yield great quantities of ground water may be found in the Hawaiian Islands and areas in the north-western United States. Although fractures may contribute significantly to the permeability of rocks near the surface of the earth, at depths greater than 200 or 300 ft the fractures generally are compressed and will not yield appreciable quantities of water to a well.

Porosity and permeability of indurated earth materials may also be affected by chemical and mechanical weathering processes, but generally not at depths greater than 50–100 ft. The degree and extent of weathering are controlled by climate, topography, time, and the chemical composition of the rock. In general, igneous and metamorphic rocks are more susceptible to chemical decomposition than are sedimentary rocks.

D. Stratigraphy, geologic structure, and topography

Stratigraphy is the branch of geology which deals with the formation, composition, sequence, and correlation of stratified earth materials. In ground-water studies it is important to know the mode of formation of a deposit and the nature of the grains which compose it, because these factors control the porosity and permeability of the deposit. The sequence of stratified deposits in natural environments of deposition is such that sediments of varied physical characteristics are often deposited in distinct layers. For example, sedimentary layers of sand and gravel commonly are found alternating with layers of silt and clay. Since the permeability of the coarse-grained deposits may be many thousand times greater than that of the fine-grained deposits, most of the water which moves through the ground is transmitted through the segregated coarse-grained layers. Finally, and probably the most significant aspect of stratigraphy relative to ground-water studies, is the use of stratigraphic correlation techniques to locate water-bearing strata. Suppose, for example, that the drilling log of a productive water well shows that drilling began in limestone, progressed through 100 ft of shale, and

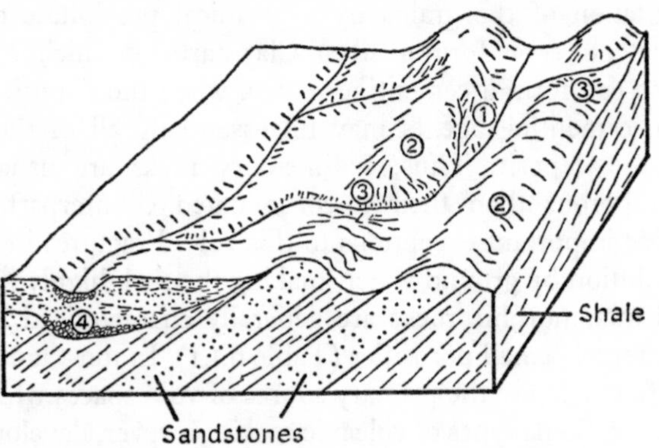

Fig. 6.1.1 Geologic factors control the occurrence of water.
Alternating strata of sandstone and shale have been tilted and erosion has cut a stream valley (1) into an exposed shale stratum. The sandstone strata, being more resistant to erosion, form the ridges (2) which parallel the valley. Where the sandstone ridges have been breached by the stream (3), the relatively permeable sandstones may receive water from the stream. The buried stream channel (4) contains gravels and other unconsolidated sediments which were deposited by a stream at an earlier time.

then water was obtained in a sandstone layer immediately beneath the shale. Several miles away, another well is to be drilled in the vicinity of a rock outcrop which exposes the contact between the limestone and the shale. The concept of stratigraphic correlation would indicate that the well would have to be drilled through approximately 100 ft of shale before the sandstone aquifer could be tapped. Thus, stratigraphy is important in ground-water studies because it helps to define the nature, location, and extent of aquifers.

Aquifers and other geologic formations may become folded or broken (faulted) because of the stresses which develop within the crust of the earth. These deformations are referred to as the geologic structure. The structure of geologic formations influences ground-water occurrence and movement because the folding and faulting of strata control the localization of areas where ground water may enter and leave each stratum. Figure 6.1.1 illustrates how different ground-water systems could result from various combinations of geologic structure and stratigraphy.

The influence of topography on the occurrence and movement of ground water can also be seen in fig. 6.1.1. Land-surface topography affects the localization of ground-water movement into and out of the ground and also controls the nature and location of surface hydrologic features, such as lakes and streams. Note in fig. 6.1.1. how surface topography is controlled by geologic structure and stratigraphy.

E. Confined (artesian) and unconfined ground water

A discussion of the effects of stratigraphy, geologic structure, and topography on ground water would not be complete without some mention of the two modes of ground-water occurrence: confined (artesian) and unconfined.

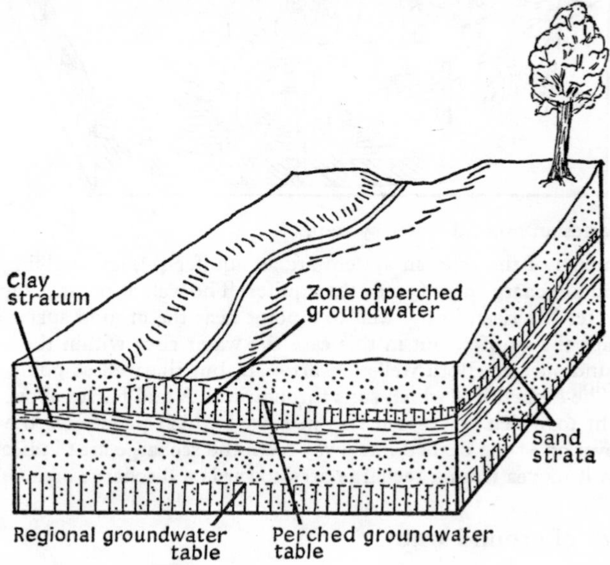

Fig. 6.1.2 Perched ground water.
The illustration shows how a zone of ground water may develop apart from the regional body of ground water. The lateral extent of a body of perched ground water is determined by the extent of the underlying impermeable stratum. The perched water in the illustration is recharged by percolation of stream water from the surface. Recharge may also occur directly from percolation of precipitation. The lower (regional) body of ground water is not recharged by percolating surface waters where it is overlain by an impermeable stratum.

Ground water which is not overlain by relatively impermeable materials is considered to be unconfined. The upper surface of a zone of unconfined ground water is called the *water-table*. By definition, the pressure of the water at every point on the water-table is equal to atmospheric pressure. In many places a zone of unconfined water may exist near the surface of the ground where downward movement of percolating water is impeded by an underlying impermeable stratum. A body of water which is isolated in this way from other ground water is termed *perched* ground water (fig. 6.1.2).

Artesian water is that which is confined beneath a relatively impermeable stratum such that, if a well penetrates the confined zone, water will rise into the well to an elevation above that of the confined zone (fig. 6.1.3). If the artesian pressure in a confined aquifer is great enough water may rise in the well to an elevation above the land surface. This phenomenon is called a flowing artesian well.

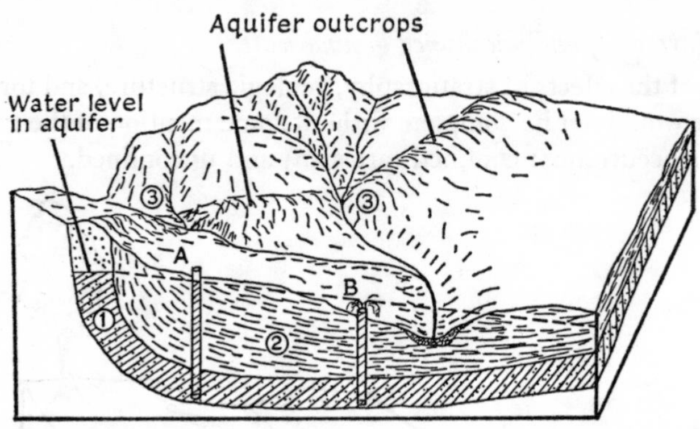

Fig. 6.1.3 An artesian ground-water system.

The basic elements of the artesian system are an aquifer (1), an overlying impermeable stratum (2), and a recharge area (3) for the aquifer. The well at point A taps the artesian aquifer, and water has risen in the well to a point near the ground surface. The well at point B also taps the aquifer, but in this case the water rises within the well to a point above the ground surface. Both wells are artesian, but the well at point B is a *flowing* artesian well. The dashed line which runs through well A and passes above well B represents the height to which water will rise in a well which penetrates the aquifer. The decrease in elevation of the dashed line away from the recharge area indicates that water loses energy as it moves from a recharge area towards a discharge area.

3. Movement of ground water

Ground water is always in motion. Movement is from a recharge area (usually where precipitation from the surface has percolated through the ground) to a discharge area (where ground water emerges from the ground in the form of a spring, seep, or discharge from a well). Because infiltration of precipitated moisture is the principal source of ground water, it is important to recognize that infiltration and percolation vary from point to point on the earth's surface. This variation is produced by the geologic and topographic factors which have been discussed in the preceding sections of this chapter.

The path of ground-water flow and the rate at which ground water moves are determined not only by the geologic conditions already discussed but also by hydraulic conditions. All things considered, water follows the path of least resistance. Movement of water from one point to another is caused by a difference in flow potential or 'head' between the points. In ground water, head usually consists of two components: a *pressure-head* component and an *elevation-head* component. We are familiar with elevation head because it provides most of the flow potential in surface-water streams. Water in stream channels invariably responds to gravity and flows from higher elevations to lower elevations. Pressure head is also no stranger. Everyone has witnessed the flow of a fluid from a point of high pressure to a point of low pressure: when air is released from a balloon or when tooth paste is squeezed from its tube.

Total flow potential or 'head' for ground water at any point in the earth is the sum of the pressure head and the elevation head at that point. How does one go about adding elevation and pressure? The pressure component of total head is measured in force per unit area (e.g., pounds per square foot); the elevation component is measured as a length (e.g., feet above sea-level). To add these two components of flow potential, it is necessary to modify one so that it will become dimensionally equivalent with the other. If pressure is divided by the unit weight (e.g., pounds/ft^3) of the fluid which is causing the pressure, the quotient has the dimensions of length.

$$P = \text{Pressure} = \frac{\text{Force}}{\text{Area}}$$

$$W = \text{Unit weight} = \frac{\text{Force (weight)}}{\text{Volume}}$$

$$\frac{P}{W} = \frac{\dfrac{\text{Force}}{\text{Area}}}{\dfrac{\text{Force}}{\text{Volume}}} = \frac{\text{Volume}}{\text{Area}} = \text{Length}$$

Therefore, if we let the elevation head be represented by the letter Z the flow potential or total head (H) at a point is given by

$$H = \frac{P}{W} + Z$$

Thus, total head is expressed as a length. This length has physical significance because it represents the height of a column of water required to produce a pressure at its base equal to the total flow potential.

If water is not moving the total head is a constant at all points in the water. For example, the flow potential at the bottom of a lake is equal to the flow potential at the surface of the lake.

Moving water loses flow potential as it moves. The stored or potential energy is transformed into heat energy because of frictional resistance to flow. Head loss in moving water is a function of the rate of flow and of the resistance to flow. In the case of ground-water flow, resistance to flow is usually represented by the coefficient of permeability, K, as given in

$$Q = KA\frac{H}{L}$$

where Q is discharge rate (gallons per minute), K is the coefficient of permeability, A is the cross-sectional area of flow, and H/L is the unit loss in total head due to flow between two points a distance L apart. This equation, basic to ground-water flow, is known as Darcy's Law. Henri Darcy was a French civil engineer who in 1856 in Dijon, France, reported his experiments on the relationship between head loss and discharge of water as it passed through the sand filters utilized in the water system of the city of Dijon. Darcy's Law states that discharge is directly proportional to head loss.

The discharge as determined by Darcy's Law is related to the actual velocity of ground-water movement. A rearrangement of the terms in Darcy's Law gives

$$\frac{Q}{A} = K\frac{H}{L} = V$$

where V is the discharge per unit area and is called a *volume flux*. Because V has the dimensions of velocity, it is commonly confused with the velocity of ground-water movement. Since V is actually a measure of discharge per unit area, the flow velocity is equal to V only in the case where the cross-sectional area A is completely open to flow. In earth materials this area includes the cross-sectional area of the mineral grains as well as the cross-sectional area of the spaces between grains. Of course, the fluid passes only through the open spaces between grains. The open spaces (porosity) usually constitute from about $\frac{1}{4}$ to $\frac{1}{2}$ of the total volume (or cross-sectional area) of a granular porous material. Thus, for unconsolidated sediments the velocity of flow is approximately two to four times the volume flux. If we represent the porosity of a sediment by the letter p velocity of ground-water movement through the sediment is given by

$$\text{Velocity} = v = \frac{KH}{pL}$$

Ground-water flow velocities in nature may vary in the extreme from several feet per second to less than a foot per year. The normal rate of flow of ground water is probably between 5 ft/yr and 5 ft/day (Todd, 1959). Thus, compared to flow rates in surface streams, motion of ground water is extremely slow.

REFERENCES

DAVIS, S. N. and DEWIEST, R. J. M. [1966], *Hydrogeology;* (John Wiley and Sons, Inc., New York), 463 p.

HEATH, R. C. and TRAINER, F. W. [1968], *Introduction to Groundwater Hydrology*; (John Wiley and Sons, Inc., New York), 284 p.

JOHNSON, A. I. [1967], Groundwater; *Transactions of the American Geophysical Union*, **48** (2), 711–24.

JOHNSON, E. E. [1966], *Groundwater and Wells;* (Edward E. Johnson, Inc., Saint Paul, Minnesota), 440 p.

MEINZER, O. E. [1923], The occurrence of groundwater in the United States, with a discussion of principles; *U.S. Geological Survey Water Supply Paper* 489, 321 p.

TODD, D. K. [1959], *Groundwater Hydrology;* (John Wiley and Sons, Inc., New York), 336 p.

7.1. Open Channel Flow

D. B. SIMONS

Department of Civil Engineering, Colorado State University

1. Introduction

Flow in open channels has been nature's way of conveying water on the surface of the earth since the beginning of time. Furthermore, these streams have constantly been the subject of study by man since he has been alternately blessed by the life-giving quality of streams under control and plagued by their destructive quality when out of control, such as in time of flood. Hence, the characteristics of rivers are of importance to everyone dealing with water resources, whether from the viewpoint of geomorphology, hydraulics, flood control, navigation, stabilization or water-resources development for municipalities, and industry.

2. Properties of fluids

The following physical properties of fluids influence fluid motion, channel geometry, and help explain sediment transport.

Mass is the amount of substance in matter measured by its resistance to the application of force.

Density is mass per unit volume, and is commonly symbolized by the Greek letter ρ (rho).

Weight is the force that gravity exerts on a mass; $W = gM$, where g is the acceleration of gravity.

Specific Weight is the weight per unit volume and is symbolized by the Greek letter γ (gamma); $\gamma = \rho g$.

Viscosity is the property of fluids that resists deformation and is commonly symbolized by the Greek letter μ (mu).

Shear is a property of fluid motion that is closely related to viscosity. It is the tangential force or stress per unit area that is transmitted through a unit thickness of a fluid. Shear, τ (tau), is related to viscosity by the equation $\tau = \mu \dfrac{dv}{dy}$.

Temperature affects the density of liquids slightly and the viscosity significantly. That is, water is essentially incompressible. The viscosity of liquids decrease with increasing temperature. Water temperature in open channels can vary as much as 40° F within a 24-hour period.

Elasticity and *Surface Tension* have little effect on flow in open channels, including sediment transport.

3. Types of flow

There are several types of flow in open channels. These include laminar flow and turbulent flow; uniform flow and non-uniform (or varied) flow; steady flow and unsteady flow; and tranquil flow, rapid flow, and ultra-rapid flow.

A. Laminar flow

Fluid motion may occur as laminar or turbulent flow. In laminar flow each fluid element moves along a specific path with a uniform velocity. There is no diffusion between the stream tubes, layers, or elements of flow; and accordingly, there is no turbulence. The energy used in maintaining viscous flow is dissipated in the form of heat from the friction within the fluid.

With laminar flow, the shear stress $\tau = \mu \dfrac{dv}{dy}$ being transmitted through each unit of depth varies uniformly from zero at the surface to a maximum at the stream bed, while the velocity curve is parabolic in shape, with its vertex at the surface.

In stream flow disturbances are present in such magnitude that laminar flow is rarely found. As velocity or depth increases, a given condition of laminar flow will reach a critical condition and become turbulent flow.

B. Reynolds' number

Values of *Reynolds' number* (*Re*) can be used to predict the type of flow. This dimensionless number includes the effects of the flow characteristics, velocity, and depth, and the fluid properties density and viscosity.

$$Re = \frac{VR\rho}{\mu} \tag{1}$$

The ratio $\dfrac{\mu}{\rho}$ is a fluid property called the kinematic viscosity, commonly designated ν (nu). Using this property,

$$Re = \frac{VR}{\nu} \tag{2}$$

With the value of Reynolds' number less than 500, laminar flow will prevail; whereas, with values in excess of 750, turbulent flow will prevail for smooth boundary conditions. For natural channels the critical value will be near 500 due to bed roughness.

The Reynolds' number is defined as

$$Re = \frac{\text{Inertia force}}{\text{Viscous force}} \tag{3}$$

That is, *Re* is an index of the relative importance of viscous forces in a hydraulic problem. Using Newton's second law of motion to define the inertial force, the expression $\tau = \mu \dfrac{dv}{dy}$ to define the viscous force and dimensional analysis

$$Re = \frac{\rho \dfrac{L^3 L}{T^2}}{\mu \dfrac{LL^2}{TL}} \tag{4}$$

substituting

$$V = \frac{L}{T} \quad \text{and} \quad L = D$$

$$Re = \frac{\rho V^2 D^2}{\mu VD} = \frac{VD\rho}{\mu} \tag{5}$$

Many other dimensionless parameters which have the same form as the foregoing Reynolds' number are utilized in the analysis of open-channel flow problems, such as: $\dfrac{wd}{\nu}$ and $\dfrac{V_* d}{\nu}$,

where w = fall velocity of sediment or bed material;

d = median diameter of the sediment or bed material;

V_* = shear velocity which is equal to $\sqrt{gRS}$.

Problem No. 1. A sheet of water 0·25 ft deep is flowing over a smooth surface at 1·0 ft/sec. Compute the Reynolds' number (Re) of the flow. Will the flow likely be laminar or turbulent? (Kinematic viscosity (μ) = 1·21 × 10⁻⁵ ft²/sec).

$$Re = \frac{(1{\cdot}0)\,(0{\cdot}25)}{1{\cdot}21}\,(10^5) = \underline{\underline{20{,}700}} \text{ Turbulent}$$

C. Froude number

The Froude number F_r is another dimensionless parameter frequently used to describe flow conditions. It is an index to the influence of gravity in flow situations where there is a liquid–gas interforce – such as in an open channel. The Froude number is usually defined as

$$F_r = \left(\frac{\text{Inertia force}}{\text{Gravity force}}\right)^{\frac{1}{2}} \tag{6}$$

or

$$F_r = \frac{\text{Velocity of flow}}{\text{Velocity of a small gravity wave in still water}} \tag{7}$$

Referring to the first definition and dimensional analysis

$$F_r = \left(\frac{\rho \dfrac{L^3 L}{T^2}}{\Delta\gamma L^3}\right)^{\frac{1}{2}} \tag{8}$$

Substituting

$$V = \frac{L}{T}$$

$$F_r = \left(\frac{\rho V^2 L^2}{\Delta\gamma L^3}\right)^{\frac{1}{2}} = \frac{V}{\sqrt{\Delta\gamma L/\rho}} \tag{9}$$

where L = a length dimension;
$\Delta\gamma$ = difference in specific weight of the fluids – usually air and water;
V = average velocity of flow;
ρ = mass density of the fluid.

In open-channel flow $\Delta\gamma$ is essentially the same as γ for water alone, since the density of air is so small. If D (depth) or hydraulic radius $R = \frac{A}{P}$ is used for the L dimension and $\frac{\gamma}{\rho}$ replaced by its equivalent g, then the Froude number becomes

$$F_r = \frac{V}{\sqrt{Dg}} \tag{10}$$

Note that in wide shallow channels depth of flow and the hydraulic radius are nearly equal.

A Froude number of unity indicates critical flow; less than unity indicates tranquil flow, the common variety of turbulent flow; and greater than unity, 'rapid flow'. In keeping with the second definition of Fr the $\sqrt{gD}$ term is the velocity at which a small wave travels in still water of depth D. For example, one can throw a pebble into a stream and, comparing the velocity of the waves caused by the pebble and the velocity of the flow, determine whether flow is sub-critical, critical, or super-critical.

Problem No. 2. Compute the Froude number of an open-channel flow where the mean velocity is 5 ft/sec and the depth is 1·5 ft. Describe the flow with respect to critical flow.

$$F_r = \frac{V}{\sqrt{gD}}$$

$$F_r = \frac{5}{\sqrt{(32{\cdot}2)(1{\cdot}5)}} = 0{\cdot}73 \ \underline{\underline{\text{tranquil or subcritical}}}$$

D. Turbulent flow

Turbulence, as a complicated pattern of eddies, produces small velocity fluctuations at random in all directions with an average time value of zero. Energy dissipation is high in turbulent flow due to the continuous interchange of finite masses of fluid between neighbouring zones of flow. The resistance to flow increases with approximately the square of the velocity.

Turbulent flow, as a result of this mixing and exchange of energy, has a more uniform distribution of velocity from top to bottom than laminar flow. The velocity distributions for laminar and turbulent flow are compared qualitatively in fig. 7.1.1.

Several parameters have been developed to describe turbulent flow. The shear stress in turbulent flow is defined as $\tau = \eta \frac{dv}{dy}$, in which η (eta) is termed the *eddy viscosity*. A parameter used to describe the magnitude of turbulent velocity fluctuations is the root-mean-square ($\sqrt{V'^2}$) of the deviations from the mean

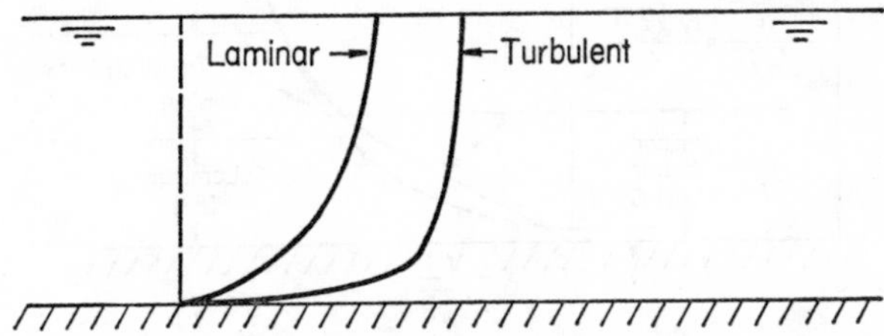

Fig. 7.1.1 Comparison of velocity distribution in laminar and turbulent flow.

velocity. The mean size of the turbulent eddies is measured by the mixing length (l) – the distance through which fluid elements move before diffusing with the surrounding fluid. The diffusion coefficient, $\epsilon = l\sqrt{V'^2}$, is a measure of the mixing process. The general pattern of variation of these parameters in turbulent flow is shown in fig. 7.1.2.

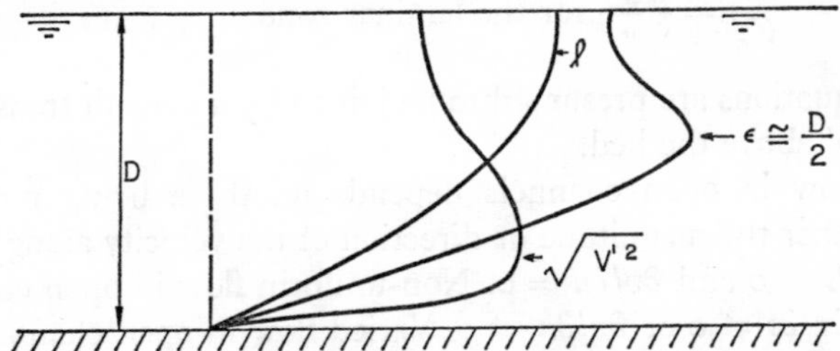

Fig. 7.1.2 Variation in mean turbulence characteristics with depth (After Rouse, 1946).

From a combination of experimental study and theory the distribution of velocity in turbulent flow over smooth boundaries has been determined. A thin layer of laminar flow persists at the boundary surfaces. This layer is called the laminar sub-layer. The theoretical velocity curve is a composite of the logarithmic turbulent flow pattern and the nearly linear laminar pattern joined by a transition curve as shown in fig. 7.1.3.

In fig. 7.1.3 δ' is the thickness of the laminar sub-layer, and it is defined by the equation

$$\delta' = \frac{11{\cdot}6\,\nu}{\sqrt{\frac{\tau}{\rho}}} = \frac{11{\cdot}6\,\nu}{V_*} \tag{11}$$

If extended towards the bed the logarithmic form of the turbulent velocity distribution will yield zero velocity at a distance y' above the bed. Experiments show that $y' = \delta'/107$.

The term $\sqrt{\frac{\tau}{\rho}}$ is a common parameter called the shear velocity (V_*), and is equal to $\sqrt{RSg}$.

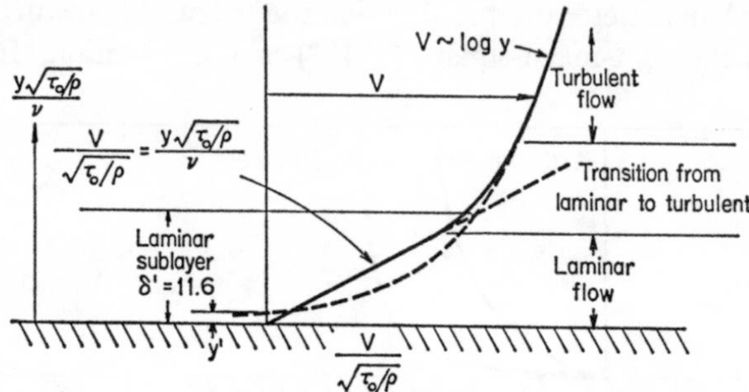

Fig. 7.1.3 Details of flow near the bed of an open channel (After Albertson, Barton and Simons, 1960).

The Karman–Prandtl equation describes the velocity distribution of turbulent flow over a smooth bed as

$$\frac{v}{V_*} = 5{\cdot}75 \log_{10} \frac{V_* y}{\nu} + 5{\cdot}5 \text{ for the turbulent zone}$$

and

$$\frac{v}{V_*} = \frac{V_* y}{\nu} \text{ for the laminar zone}$$

The two equations are presumed to be joined by a smooth transition curve at the distance δ' above the bed.

Uniform flow in open channels depends on there being no change with distance in either the magnitude or direction of the velocity along a stream line, i.e. both $\partial v/\partial s = 0$ and $\partial v/\partial n = 0$. Non-uniform flow in open channels occurs when either $\partial v/\partial s \neq 0$ or $\partial v/\partial n \neq 0$. Varied flow in open channels is a type of non-uniform flow which occurs when $\partial v/\partial s \neq 0$. Steady flow occurs when the velocity at a point does not change with time, i.e., $\partial v/\partial t = 0$. When the flow is unsteady $\partial v/\partial t \neq 0$. An example of unsteady flow is a flood wave or a travelling surge.

Unlike laminar and turbulent flow, tranquil flow and rapid flow exist only with a free surface or inner face. The criterion for tranquil and rapid flow is the Froude number $F_r = V/(gD)^{1/2}$. When $F_r = 1{\cdot}0$ the flow is critical; when $F_r < 1{\cdot}0$ the flow is tranquil; and when $F_r > 1{\cdot}0$ the flow is rapid. Ultra-rapid flow involves slugs or waves superposed over the uniform flow pattern, which makes the flow both non-uniform and unsteady.

Uniform flow in an open channel occurs with either a mild, a critical, or a steep slope. With a mild slope the flow is tranquil; with a critical slope the flow is critical; and with a steep slope the flow is rapid.

4. Velocity distribution over rough beds

Stream channels have rough beds. The roughness is expressed in terms of K_s, which is equivalent to the diameter of the sediment grains which compose the bed. The dimension of K_s is larger than that of δ', and therefore the sub-layer ceases to exist for practical purposes. Turbulent flow is assumed to occur throughout the depth. The Karman–Prandtl velocity equation for rough beds is

$$\frac{v}{V_*} = \frac{2{\cdot}303}{\kappa} \log_{10} \frac{y}{\kappa_s} + 8{\cdot}5 \qquad (12)$$

where kappa (κ) is the so-called universal velocity coefficient, which is approximately 0·4 for fixed boundary channels. The distribution of velocity in accordance with this equation is a straight line when plotted on semi-log paper with a slope of $\dfrac{\kappa}{2{\cdot}303\, V_*}$.

5. Velocity and discharge measurements

Velocity is a vector quantity, hence both its direction and magnitude must be measured. The discharge in an open channel or pipe, in the most simple terms, is the product of area and average velocity measured normal to the area.

A. Current meters

In an open channel one of the most common methods of measuring discharge involves integration of the velocity distribution across the flow section using a current meter, pitot tube, or similar device. The current meter consists of an instrument with an impeller mounted on a rod or cable. If a cable is used to suspend the meter in the flow there must be a streamlined weight at its lower end, below the current meter, of sufficient magnitude to overcome the force of the stream, enabling the operator to place the meter at any desired point in the vertical. Having taken sufficient point measurements in a vertical to establish the average velocity, the operator moves to a new vertical, by wading in small streams or by cable car or boat on large streams.

The average velocity in a vertical is located at approximately 0·6 depth below the surface and can be more precisely determined by averaging the point velocities at 0·2 and 0·8 depth. In flows with a depth less than 1·5 ft a single-point measurement is taken in each vertical in the stream cross-section at 0·6 depth. In deeper streams the 0·2 and 0·8 measurements are taken and averaged in each vertical. Discharge determined by the 0·2 and 0·8 measurements is illustrated in the example problem.

Problem No. 3. The computation of stream discharge based on current meter measurements is illustrated in Table 7.1.1.

Similarly, floats can be used to estimate the magnitude and direction of surface velocities. Greater accuracy can be achieved by using a submerged or partly submerged float which measures the velocity at more nearly 0·6 depth. Also, it is more independent of the effect of wind and waves, but may be bothered by debris.

TABLE 7.1.1

			Velocity						
Distance from bank	*Depth*	*Observation depth*	*At Point*	*Mean in vertical*	*Mean in section*	*Area*	*Mean depth*	*Width*	*Discharge (q)*
0	0	0	0	0					
					0·78	1·70	0·85	2	1·33
2	1·70	0·35	1·52	1·56					
		1·35	1·60						
					1·73	4·40	2·20	2	7·61
4	2·70	0·54	1·91	1·89					
		2·16	1·88						
					2·08	13·80	3·45	4	28·7
8	4·20	0·84	2·35	2·27					
		3·36	2·19						
					2·33	8·60	4·30	2	20·1
10	4·40	0·88	2·41	2·38					
		3·52	2·34						
					2·37	8·40	4·20	2	19·94
12	4·00	0·80	2·31	2·36					
		3·20	2·40						
					2·05	13·60	3·40	4	27·9
16	2·80	0·56	1·92	1·74					
		2·24	1·57						
					1·49	4·30	2·15	2	6·41
18	1·50	0·30	1·35	1·24					
		1·20	1·13						
					0·62	1·50	0·75	2	0·93
20	0	0	0	0					

$Q = \Sigma q = 112{\cdot}9$ cfs.

B. *Dye-dilution*

Various fluorescent tracer-type dyes can be detected and accurately measured at very low concentrations using a fluorometer. This makes it possible to successfully measure discharge by various dye-dilution techniques. Two methods can be used. One involves

$$Q = q\frac{C_1}{C_2} \qquad (13)$$

where q is the injected discharge, C_1 is concentration of the dye in the injected flow, and C_2 is the concentration of dye in the unknown discharge Q. On small streams injection of a small steady q at known concentration C_1 for about 15 minutes will enable C_2 to stabilize. Only C_2 at its plateau needs to be determined to compute the stream discharge Q since the effect of q is small.

The second method is called the total recovery method. The discharge is evaluated using the relation

$$Q = \frac{(\text{Vol. of dye})(\text{Conc. of dye})}{\int_0^\infty C\,dt} \tag{14}$$

The integral term is the total area under the concentration–time curve, where C is the measured dye concentration at time t at the point of sampling. In both relations any material background fluorescence must be considered in measuring effective dy concentrations.

C. Weirs

The weir is extensively used to measure flow in open channels. It is essentially an overflow structure extending across the channel normal to the direction of flow, see fig. 7.1.4.

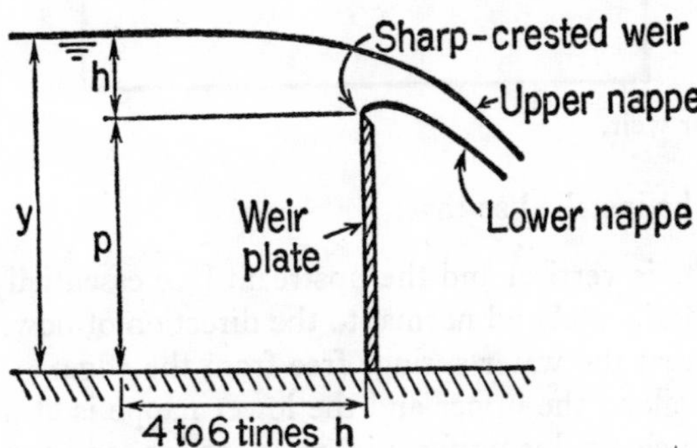

Fig. 7.1.4 Sharp-crested weir.

Weirs are classified according to shape. The most common ones are the standard uncontracted weir, also known as the suppressed weir, the contracted weir, the V-notch weir, the trapezoidal weir, and the broad-crested weir. The first four are sharp crested, as shown in fig. 7.1.4.

Many formulae have been suggested by various experimenters, but only a few are presented.

In 1823 Francis suggested the equation

$$Q = 3{\cdot}33 L h^{3/2} \tag{15}$$

for uncontracted (suppressed) weirs in which L is the length of crest in feet and h is the head on the weir in feet. For the contracted weir

$$Q = 3{\cdot}33 \left(L - \frac{nh}{10}\right) h^{3/2} \tag{16}$$

where n is the number of horizontal end contractors with a simple weir only contracted at the sides of the channel $n = 2$.

For a triangular weir $$dQ = C_d x\sqrt{2gy}\,dy \tag{17}$$

from which $$Q = \frac{8}{15}\,C_d\sqrt{2g}\,\tan\frac{\theta}{2}\,h^{5/2} \tag{18}$$

see fig. 7.1.5.

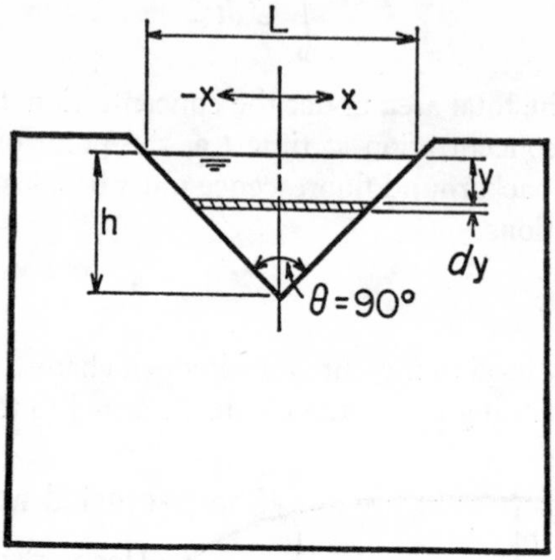

Fig. 7.1.5 Triangular weir.

The weirs should be installed so that:

1. The weir plate is vertical and the upstream face essentially smooth.
2. The crest is horizontal and normal to the direction of flow. The crest must be sharp, so that the water springs free from the edge.
3. The pressure along the upper and the lower nappe is atmospheric.
4. The approach channel is uniform in cross-section, and the water surface is free of surface waves.
5. The sides of the channel are vertical and smooth, and they extend downstream from the crest of the weir.

in order for the equations to apply.

Still greater accuracy can be achieved with all types of measuring devices if they are calibrated after construction.

Many other methods of measuring discharge in open channels and pipes are utilized. The principles utilized are (1) volumetric, (2) use of the Bernoulli momentum and continuity equations, (3) drag on an object in the flow, (4) mass flux measurements, and (5) by the slope area method. The measuring devices include: Venturi meters, Parshall flumes, nozzles, orifices, gates, weirs, spillways, contracted openings, vane meters, rotometers, and wobbling discs. Refer to any fluid mechanics, such as Albertson, Barton, and Simons [1960], for further details.

6. Hydraulic and energy gradients

The hydraulic gradient in open-channel flow is the water surface. The energy gradient is above the hydraulic gradient a distance equal to the velocity head.

The fall of the energy gradient for a given length of channel represents the loss of energy, either from friction or from friction and other influences. The relationship of the energy gradient to the hydraulic gradient reflects not only the loss of energy but also the conversion between potential and kinetic energy. For uniform flow the gradients are parallel and the slope of the water surface represents the friction-loss gradient. In accelerated flow the hydraulic gradient is steeper than the energy gradient, indicating a progressive conversion from potential to kinetic energy. In retarded flow the energy gradient is steeper than the hydraulic gradient, indicating a conversion from kinetic to potential energy. The Bernoulli theorem defines the progressive relationships of these energy gradients.

For a given reach of channel ΔL, the average slope of the energy gradient is $\Delta h_L/\Delta L$, where Δh_L is the cumulative loss through the reach. If these losses are solely from friction, Δh_L will become Δh_f and

$$\Delta h_f = \frac{S_2 + S_1}{2} \Delta L \tag{19}$$

where S_1 and S_2 are the slopes of the energy gradient at the ends of reach ΔL.

7. Energy and head

If streamlines of flow in an open channel are parallel and velocities at all points in a cross-section are equal to the mean velocity V the energy possessed by the water is made up of kinetic energy and potential energy. Referring to fig. 7.1.6,

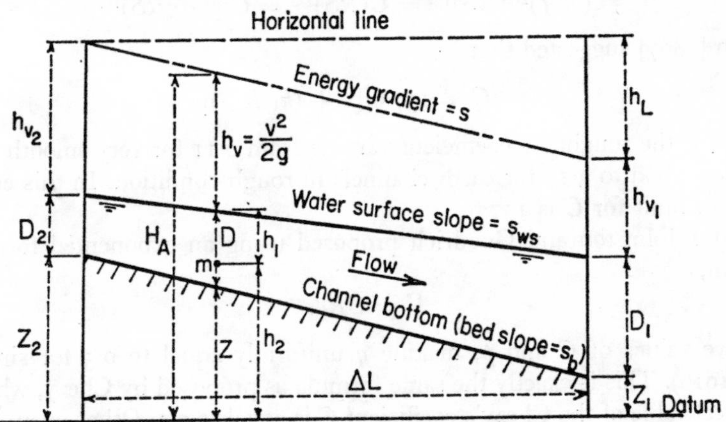

Fig. 7.1.6 Characteristics of open-channel flow.

the potential energy of mass M is $\gamma(h_1 + h_2)$ and the kinetic energy of M is $\gamma \frac{V^2}{2g}$. Hence, the total energy of each mass particle is:

$$E_m = \gamma h_1 + h_2 + \frac{V^2}{2g} \tag{20}$$

Applying the above relationship to the total discharge Q in terms of the unit weight of water γ,

$$E = Q\gamma\left(D + Z + \frac{V^2}{2g}\right) \qquad (21)$$

where E is total energy per second at the cross-section.

The parentheses term in equation (21) is the absolute head:

$$H_A = D + Z + \frac{V^2}{2g} \qquad (22)$$

Equation (22) is the Bernoulli Equation.

The energy in the cross-section referred to the bottom of the channel is termed the specific energy. The corresponding head is referred to as the specific energy head, and is expressed as:

$$H_E = D + \frac{V^2}{2g} \qquad (23)$$

Where $Q = AV$, equation (23) can be stated:

$$H_E = D + \frac{Q^2}{2ga^2} \qquad (24)$$

8. Flow equations

One of the common open-channel flow equations for estimating average velocity is the Chezy equation

$$V = (8g/f)^{1/2}(RS)^{1/2} = C(RS)^{1/2} = C/g^{1/2}(gRS)^{1/2} \qquad (25)$$

Bazin [1897] suggested that

$$C = 157{\cdot}6/[1 + (k_1/R^{1/2})] \qquad (26)$$

where k_1 is the roughness coefficient varying from 0·11 for very smooth cement or planed wood to 3·17 for earth channels in rough condition. In this equation the upper limit for C is 157·6.

In 1911 Johnston and Goodrich proposed using an exponential formula of the form,

$$V = CR^p s^q \qquad (27)$$

and gave values of C and p, making q uniformly equal to 0·5 for simplicity (Ellis, 1916). This is exactly the same formula as proposed by Chezy, where the numerical value of the Chezy's coefficient C is equal to 0·5. Other open-channel flow equations that are often used are given by N. G. Bhowmik [1965] and Garbrecht [1961].

In an effort to correlate and systematize existing data from natural and artificial channels, Manning [1889] proposed the equation

$$V = (1{\cdot}5/n)R^{2/3}\,S^{1/2} \qquad (28)$$

or

$$Q = AV = A(1{\cdot}5/n)R^{2/3}\,S^{1/2} \qquad (29)$$

in which n is the Manning roughness coefficient which has the dimensions of $L^{1/6}$. By comparing equation (25) with equation (28), the Chezy discharge coefficient C can be expressed as follows:

$$C = 1{\cdot}5(R^{1/6}/n) \tag{30}$$

and is related to the Manning coefficient n and the hydraulic radius $R = \dfrac{A}{P}$. The Manning n was developed empirically as a coefficient which remained a constant for a given boundary condition, regardless of slope of channel, size of channel, or depth of flow. As a matter of fact, however, each of these factors causes n to vary to some extent. In other words, the Reynolds' number, the shape of the channel, and the relative roughness have an influence on the magnitude of Manning's n. Furthermore, for a given alluvial bed of an open channel the size, pattern, and spacing of the sand waves vary, so that n varies. Despite the shortcomings of the Manning roughness coefficient, it is used extensively.

The magnitude of Manning roughness is given in Table 7.1.2 for rigid channels

TABLE 7.1.2 Manning roughness coefficients for various boundaries

Boundary	*Manning Roughness* n *in* $(ft)^{1/6}$
Very smooth surfaces such as glass, plastic, or brass	0·010
Very smooth concrete and planed timber	0·011
Smooth concrete	0·012
Ordinary concrete lining	0·013
Good wood	0·014
Vitrified clay	0·015
Shot concrete, untrowelled, and earth channels in best condition	0·017
Straight unlined earth canals in good condition	0·020
Rivers and earth canals in fair condition – some growth	0·025
Winding natural streams and canals in poor condition – considerable moss growth	0·035
Mountain streams with rocky beds and rivers with variable sections and some vegetation along banks	0·040–0·050
Alluvial channels, sand bed, no vegetation	
1. Tranquil flow, $F_r < 1$	
Plane bed	0·014–0·02
Ripples	0·018–0·028
Dunes	0·018–0·035
Washed-out dunes or transition	0·014–0·024
Plane bed	0·012–0·015
2. Rapid flow $F_r > 1$	
Standing waves	0·011–0·015
Anti-dunes	0·012–0·020

and alluvial channels. Considering alluvial channels, note that as the form of bed roughness changes from dunes through transition to plane bed or standing waves the magnitude of Manning n decreases by approximately 50%.

9. Natural channels

The natural shape of an open channel may be markedly different from rectangles and trapezoids. However, it is usually possible to break down the complex shape of a natural open channel into simple elementary shapes for analysis. Consider fig. 7.1.7, in which flow is occurring not only in the main channel but also in the

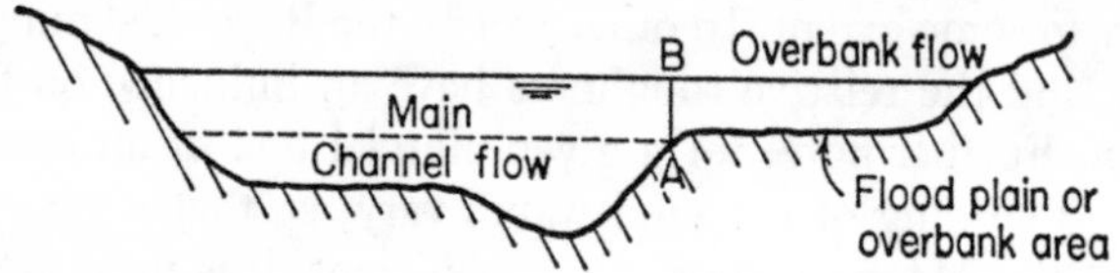

Fig. 7.1.7 Shape of natural channels.

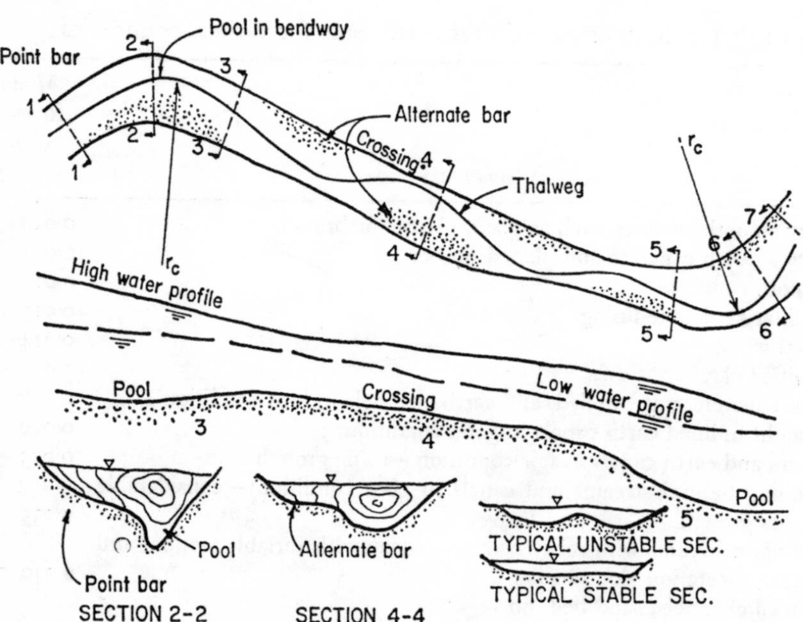

Fig. 7.1.8 Geometry of rivers.

overbank or floodplain area. In this case the hydraulic radius R, which would be obtained by using the area and the wetted perimeter for the entire section, would not be truly representative of the flow. Furthermore, the roughness coefficient in the overbank area is usually different from the coefficient in the main channel. Therefore, such a section should be divided along AB and treated as two separate sections. The plane AB, however, is not considered as a part of the wetted perimenter, since there is no appreciable shear in this plane.

Along a natural channel there are frequently pools and riffles or rapids. At low to moderate discharges the slope of water surface is relatively flat over the pools and steep over the riffles. With a further increase in stage, this condition may be reversed as for the Mississippi River (fig. 7.1.8). Therefore care must be taken in studies of natural streams to consider the correct slope for the particular discharge and reach of stream in question.

10. Forms of bed roughness and resistance to flow in alluvial channels

The primary variables which affect the form of bed roughness and resistance to flow in sand-bed alluvial channels (Simons and Richardson, 1962, 1963) include:

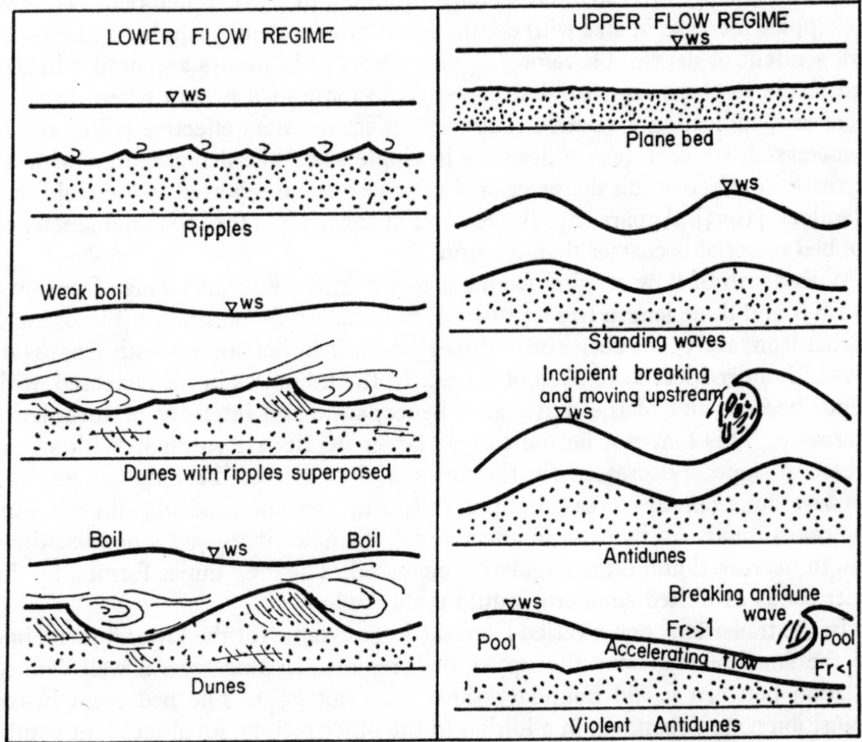

Fig. 7.1.9 Forms of bed roughness in alluvial channels (the term 'flat bed' is now preferred to 'plane bed').

the slope of the energy grade line, depth, physical size of the bed material as related to grain roughness, and fall velocity or effective median fall diameter as related to form resistance. The fall velocity or effective median fall diameter depends on the viscosity and mass density of the water sediment mixture and the mass density, size, and shape of the bed material. It reflects the principal viscous effect on flow in alluvial channels when *Re* is large. The effective median fall diameter is defined as the diameter of a sphere having a specific gravity of 2·65 and a fall velocity in distilled water of infinite extent at a temperature of

24° C equal to the fall velocity of the particle falling alone in any quiescent stream fluid at stream temperature (Haushild *et al.*, 1961; Simons *et al.*, 1963).

The regimes of flow and various forms of boundary roughness (Simons and Richardson, 1963) which can occur in alluvial channels are illustrated in fig. 7.1.9.

In the lower flow regime flow is tranquil and the water surface undulations are out of phase with the bed undulations. Resistance to flow is large, because separation of the flow from the boundary generates large-scale turbulence that dissipates considerable energy.

With depths of flow ranging from 0·4 to 1·0 ft, ripple heights range from 0·01 to 0·1 ft and length (crest-to-crest) range from 0·5 to 1·5 ft. When depth is small the ripples increase in size with depth, but at greater depths ripple size becomes independent of depth. Therefore, ripples observed in flumes are similar in size and shape to those in natural streams. A decrease in n occurs when depth is increased, indicating a relative roughness effect or when effective fall diameter is increased, which causes a decrease in ripple size. The decrease in n with an increase in effective fall diameter is similar to change reported in Leopold and Maddock [1953]. Apparently, ripples do not form when the median diameter of the bed material is coarser than 0·7 mm.

With depth of flow ranging from 0·4 to 1·0 ft, dune heights range from 0·15 to 1·0 ft and dune length from 4 to 20 ft, based upon flume studies (Simons and Richardson, 1963). In deep rivers dunes 30–60 ft in height and with lengths of several hundred feet have been observed. In the flume studies n increased with depth because size of the dunes and, hence, scale and intensity of turbulence increased. This may not be the case as larger depths are studied. With an increase in slope, n decreased for the fine sand but increased for the coarse sand because dune length increased appreciably with the fine sand but did not with the coarse sand. An increase in effective fall diameter increased n because dune length decreased and dune angularity increased. The long dunes formed by the finer sands exhibited smaller n values than ripples.

In the transition zone n varied from the largest value for the lower flow regime to the smallest value for the upper flow regime. In this zone a well-defined relation between n and boundary shear does not exist. The bed form in the transition zone depended, in addition to the other factors, on antecedent conditions. Starting with dunes, slope and/or depth could be increased to relatively large values before plane bed or standing waves occurred. Conversely, with a plane bed and/or standing waves, the slope and/or depth could be decreased to relatively small values before dunes developed.

In the upper flow regime n values are small because surface or grain resistance predominates. However, the energy dissipated by the wave formation with the standing waves and the formation and breaking of the waves with antidunes increases n. Standing waves and antidunes, from the standpoint of wave mechanics, are rapid flow phenomena.

Standing waves are sinusoidal, in-phase sand and water waves (fig. 7.1.9) that build up in amplitude from a plane bed and water surface and gradually fade away. In the flume studies with depth of flow ranging from 0·2 to 0·6 ft the water wave height (trough-to-crest) ranged from 0·01 to 0·6 ft and was 1·5–2 times the height of corresponding sand waves. Spacing of the waves was from 2 to 5 ft. Both height and spacing of the waves increased with depth. Resistance to flow for standing waves was larger than for a plane bed and, as with a plane bed, increased with an increase in sand size. Standing waves did not occur using the two finer sands, because the mobility of the particles (effective fall diameter) allowed the development of antidunes whenever the Froude number equalled one.

Antidunes are similar to standing waves, except they increase in amplitude until they break. Breaking antidunes are similar to the hydraulic jump. The breaking wave dissipates a large amount of energy that is reflected by increased n. The increase of n is in direct proportion to the amount of antidune activity and the portion of the flume or channel occupied by the antidunes. Antidune activity increases with a decrease in effective fall diameter or with an increase in slope.

11. Comparison of flume and field phenomena

The preceding comments were based on observations of bed configurations and flow phenomena that occurred in the flume experiments and in natural rivers (field conditions) and are equally true for both situations. However, there are major differences between flume and field conditions. In the usual flumes only a limited range of depth and discharge can be investigated, but slope and velocity can be varied within a wide range. In the field a larger range of depth and discharge is common, but slope of a particular channel reach is virtually constant. Larger Froude numbers ($V/\sqrt{gD}$) can be achieved in flume studies than will occur in most natural alluvial channels because natural banks cannot withstand prolonged high-velocity flow without eroding. This erosion increases the cross-sectional area, and this causes a reduction in the average velocity and the Froude number. Rarely does a Froude number, based on average velocity and depth, exceed unity for any extended time period in a natural stream with erodible banks. In fact, rarely are natural channels truly stable when $F_r > 0{\cdot}25$. In the field, where the slope of the energy grade line is constant, the Froude number is also constant unless there is a change in the resistance to flow,

$$\left(F_r = \frac{C}{\sqrt{g}}\sqrt{S}\right)$$

Flow is more nearly two-dimensional in flumes than in natural streams. However, the main current meanders from side to side in a large flume, as it does in the field, and bars of small amplitude but large area develop in an alternating pattern adjacent to the walls of the flume. (It is on these bars that the bed forms shown in fig. 7.1.9 superpose themselves.) If very large width–depth ratios are maintained by keeping depth of flow shallow these bars may grow to the water surface.

In the field it is even more obvious that the flow meanders between the parallel banks of a straight channel, and the alternate bars which form opposite the apex of the meanders are easier to distinguish. As in a flume, if the depth is decreased the alternate bars increase in amplitude until they are close to the water surface, or even exposed. In fact, scour in the main current adjacent to a large bar may cause the water surface there to drop slightly so that the top of the bar is exposed. This has been observed in the Rio Grande and in other natural channels. If the banks are not stable erosion occurs where the high-velocity water impinges, and deposition occurs on the opposite bank. The ultimate development is a meandering stream if other factors such as slope, discharge, and size of bed material are compatible.

The alluvial bars which form on the bed of an alluvial channel is a type of roughness element that plays a very significant role in river mechanics and channel geometry. These bars are much larger than the ripples and dunes illustrated in fig. 7.1.9. Their amplitude ranges from very small to as large as the average depth of flow. Their widths range from $\frac{1}{2}$ to nearly $\frac{1}{3}$ channel width, depending upon the size of the system, and they may be several hundred feet long.

The three most common and accepted types of bars are: point bars, alternate bars, and middle bars (see fig. 7.1.8).

The position, shape, and magnitude of the alternate bars is a function of channel alignment, bed material, and width–depth ratio. These bars normally occur in the straight reaches or crossings between two consecutive bends. Conditions in the upstream bend usually dictate on which side of the channel the first alternate bar will form. Thereafter the sequence is fixed, because the second bar must be on the opposite side of the channel, and so on. In the next bend downstream the secondary circulation developed within it is usually, unless the radius of the bend is very large, of sufficient magnitude to terminate the sequence of alternate bars. Within each bend the pool forms adjacent to the outside bank and a point bar forms on the inside bank. Consequently, the alternate bars are in a sense locked in between the two bends, and their physical characteristics vary with the type of bed material, the width–depth ratio of the channel, and the characteristics of the bends. The number of bars between two bends may increase as the discharge decreases, and vice versa. The flow meanders around these alternate bars. In fact, this flow phenomena is probably the way meanders are initiated and develop with time. For example, consider the successive stages of the development of a meander in sand. Initially the channel may be straight, but the meandering of the thalweg rapidly initiates the development of the meander. Another significant point is that the wavelength of the meander is essentially constant for constant discharge throughout the development of the meander, even though the amplitude increases significantly. The rate of movement of the alternate bars is mostly just a change in size and shape as flow conditions change and the bars alter their geometry to suit the new flow conditions. However, if the bends of the channel are moving, then this allows an additional freedom of movement for the bars. Leopold [1964] has documented that bars move very slowly – of the order of a foot or so per day. Simons has verified the foregoing information on bars by observing their development and movement with time in canals and rivers, and has studied their detailed movement in the sand-bed flumes at Colorado State University. He observed that as a specific weight of the bed material decreases the rate of both change of shape and movement downstream rapidly increases.

Various methods for predicting form of bed roughness have been developed. One of the more useful methods was presented by Simons and Richardson [1963].

12. Tractive force

The tractive-force theory is formulated on the basis that stability of bank and bed material is a function of the ability of the bank and bed to resist erosion resulting from the tractive force exerted on them by the moving water.

Consider the free body of a segment of the full width of the channel as shown in fig. 7.1.10.

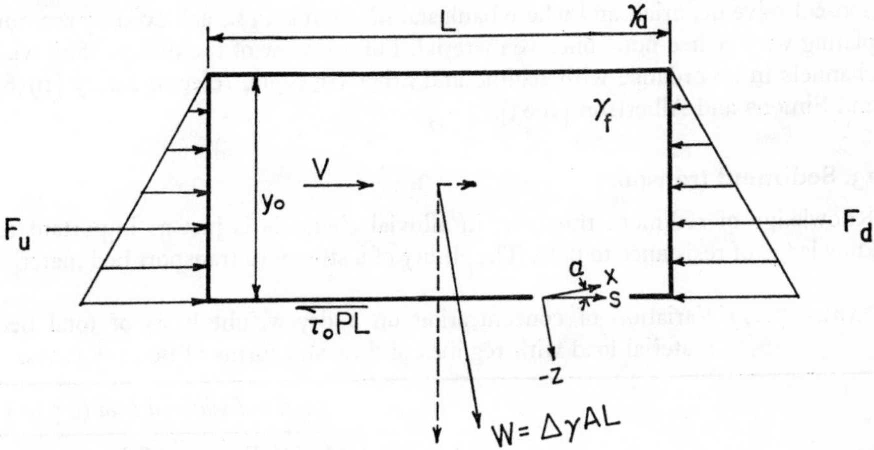

Fig. 7.1.10 Free-body diagram of segment of open-channel flow.

Equating the forces parallel to the flow yields

$$F_u + W \sin \alpha = F_d + \tau_0 p L \tag{31}$$

where W is the weight of the entire segment of fluid;

p is the wetted perimeter – that is, the length of the cross-sectional boundary which is in contact with the fluid flowing in the channel;

F_u and F_d are the upstream and downstream hydrostatic forces acting on the free body where $p = \gamma y$ – since the flow is uniform, $F_u = F_d$;

τ_0 is the average boundary shear which is retarding the flow;

A is the cross-sectional area of the flow;

L is the length of the free body segment;

α is the angle which the channel slope makes with the horizontal; and

$\Delta\gamma$ is the difference between the specific weight γ of the fluid flowing and the specific weight γ_a of the ambient fluid, normally the air.

The product $\Delta\gamma AL$ may be substituted for the weight W and the equation rearranged to solve for the boundary shear

$$\tau_0 = \Delta\gamma \frac{A}{wp} \sin\alpha - \Delta\gamma\, RS = \gamma\, RS \tag{32}$$

where R is called the hydraulic radius which is the area A divided by the wetted perimeter p; and

S is $\sin\alpha = dz/ds$, which for relatively flat slopes may be considered more conveniently as $\tan\alpha = dz/dx$.

Equation (31), it may be noted, evaluates the boundary shear in terms of the static characteristics of the geometry and the fluid.

A tractive force theory was clearly presented and illustrated by Lane [1953] to assist with the design of channels for conveying essentially clear water in coarse non-cohesive materials and where bank stabilization is to be achieved by armour plating with coarse non-cohesive material. For a review of the design of alluvial channels in accordance with regime and other concepts, refer to Lacey [1958] and Simons and Albertson [1963].

13. Sediment transport

Knowledge of sediment transport in alluvial channels is just as important as knowledge of resistance to flow. The ability of a stream to transport bed material is relatively small when the form of bed roughness consists of ripples and/or dunes. In the upper regime of flow the streams are capable of carrying much larger volumes of sediment per unit volume of water (see qualitative data in Table 7.1.3 suggested by Simons and Richardson [1963]). Some of the more

TABLE 7.1.3 Variation of concentration on a dry-weight basis of total bed material load with regimes of flow and forms of bed roughness

		Total bed material load (p.p.m.)	
Regime of flow	*Forms of bed roughness*	*Median diameter of bed material* 0·28 *mm*	*Median diameter of bed material* 0·45 *mm*
Lower flow regime	Ripples	1–150	1–100
	Dunes	150–800	100–1,200
Transition	Zone in which dunes are reducing in amplitude with increasing shear stress	1,000–2,400	1,400–4,000
Upper flow regime	Plane	1,500–3,100	—
	Standing waves	3,000–6,000	4,000–7,000
	Antidunes	5,000–42,000	6,000–15,000

useful concepts for estimating bed material discharge have been presented by Einstein [1950], Colby and Hembree [1955], Colby [1964], Simons *et al.* [1965], and Bishop *et al.* [1965]. For a more detailed treatment of the sediment problems encountered in designing and operating irrigation canals constructed in alluvium, refer to Simons and Miller [1966].

14. Summary

The basic concepts of fluid mechanics applicable to flow in open channels has been presented. For a more detailed treatment of hydraulics and fluid mechanics, see Albertson *et al.* [1960] and Albertson and Simons [1964] and other fluid-mechanics texts. Also, many valuable concepts pertinent to the design of hydraulic structures associated with the conveyance and distribution of water have been presented by the U.S. Bureau of Reclamation [1960, 1963, 1964].

REFERENCES

ALBERTSON, M. L., BARTON, J. R., and SIMONS, D. B. [1960], *Fluid Mechanics for Engineers;* (Prentice Hall, Englewood Cliffs, New Jersey), 568 p.

ALBERTSON, M. L. and SIMONS, D. B. [1964], Fluid Mechanics; In Chow, V. T., Editor, *Handbook of Applied Hydrology*, (McGraw-Hill, New York), Chapter 7, 49 p.

BAZIN, H. [1897], Etude d'une nouvelle formule pour calculer le debit des canaux decouverts; *Annales des Ponts et Chaussees*, Memoire No. 41, Vol. 14, Ser. 7, 4me trimestre, pp. 20–70.

BHOWMIK, N. G. [1965], *The Hydraulic Design of Large Concrete-lined Canals;* Thesis. Colorado State University, Fort Collins, Colorado.

BISHOP, A. A., SIMONS, D. B., and RICHARDSON, E. V. [1965], Total bed-material transport; *Proceedings of the American Society of Civil Engineers, Journal of the Hydraulics Division* 91 (HY2), 175–91.

COLBY, B. R. [1964], Discharge of sands and mean-velocity relationships in sandbed streams; *U.S. Geological Survey Professional Paper* 462-A, 47 p.

COLBY, B. R. and HEMBREE, C. H. [1955], Computations of total sediment discharge, Niobrara River near Cody, Nebraska; *U.S. Geological Survey Water Supply Paper* 1357, 187 p.

EINSTEIN, H. A. [1950], The bed-load function for sediment transportation in open channel flows; *U.S. Department of Agriculture Technical Bulletin* 1026; 71 p.

ELLIS, G. H. [1916], The flow of water in irrigation canals; *Transactions of the American Society of Civil Engineers*, Paper no. 1373, 1644–88.

GARBRECHT, G. [1961], Flow calculations for rivers and channels; Die Wasser-Wirtschaft, (Stuttgart), Parts I & II, 40–5 and 72–7. (U.S. Bureau of Reclamation Translation 402.)

HAUSHILD, W. L., SIMONS, D. B., and RICHARDSON, E. V. [1961], The significance of fall velocity and effective diameter of bed materials; *U.S. Geological Survey Professional Paper* 424-D, 17–20.

LACEY, G. [1958], Flow in alluvial channels with sandy mobile beds; *Proceedings of the Institution of Civil Engineers*, **9**, 145–64.

LANE, E. W. [1953], Progress report on studies on the design of stable channels by the Bureau of Reclamation; *Proceedings of the American Society of Civil Engineers*, **79**, 1–31.

LEOPOLD, L. B. and EMMETT, W. W., [1963], *Downstream pattern of River-bed scour and fill;* U.S. Geological Survey paper prepared for presentation of the Federal Interagency Sedimentation Conference, Jackson, Mississippi.

LEOPOLD, L. B. and MADDOCK, T. [1953], The hydraulic geometry of stream channels and some physiographic implications; *U.S. Geological Survey Professional Paper* 252, 56 p.

MANNING, R. [1889], On the flow of water in open channels and pipes; *Transactions of the Institution of Civil Engineering of Ireland*, **20**, 161–207. (Supplement, 1895, **25**, 179–207).

SIMONS, D. B. and ALBERTSON, M. L. [1963], Uniform water conveyance channels in alluvial material; *Transactions of the American Society of Civil Engineers*, **128**, 65–106.

SIMONS, D. B. and MILLER, C. R. [1966], Sediment discharge in irrigation canals; *Proceedings of the International Committee on Irrigation and Drainage, 6th Congress*, (New Delhi, India) Quest 20, Rep. 12, 20275–307.

SIMONS, D. B. and RICHARDSON, E. V. [1962], Resistance to flow in alluvial channels; *Transactions of the American Society of Civil Engineers*, **127**, 927–52.

SIMONS, D. B. and RICHARDSON, E. V. [1963], Forms of bed roughness in alluvial channels; *Transactions of the American Society of Civil Engineers*, **128**, 284–302.

SIMONS, D. B., RICHARDSON, E. V., and HAUSHILD, W. L. [1963], Some effects of fine sediment on flow phenomena. *U.S. Geological Survey Water Supply Paper*, 1498-G, 46 p.

SIMONS, D. B., RICHARDSON, E. V., and NORDIN, C. F. [1965], Bedload equation for ripples and dunes; *U.S. Geological Survey Professional Paper*, 462-H, 9 p.

U.S. BUREAU OF RECLAMATION [1960], *Design of Small Dams* (U.S. Government Printing Office, Washington, D.C.), 611 p.

U.S. BUREAU OF RECLAMATION [1960], *Design of Stable Channels with Tractive Forces and Competent Bottom Velocity;* Sedimentation Section, Hydrology Branch, Bureau of Reclamation, Denver Federal Center, Denver, Colorado.

U.S. BUREAU OF RECLAMATION [1963], *Hydraulic Design of Stilling Basins and Energy Dissipators* (Supplement of Documents, Washington, D.C.), Engineering Monographs **25**, 114 p.

U.S. BUREAU OF RECLAMATION [1964], *Design Standards No. 3: Canals and Related Structures;* Commissioner's Office. Denver Federal Center, Denver, Colorado.

8.I. The Hydrology of Snow and Ice

MELVIN G. MARCUS
Department of Geography, University of Michigan

1. Introduction

Snow and ice are significant elements of the world hydrological system, which occur subject to tremendous variations in space and time. Snow or ice are present in the atmosphere, in lakes and rivers and oceans, on the land, and even beneath the earth's surface. Sometimes their appearance is brief and local, as in the sudden snow flurry which coats the earth with a quickly melted veneer of white, or the violent hailstorm that brings to heated, summer landscapes a contradictory deluge of ice. In other places snow and ice dominate the earth's surface. The ice caps of Antarctica and Greenland and the frozen floes of the Arctic Ocean seem, from the human view at least, hostile and permanent features of our world.

Yet even the great ice sheets are transitory features when viewed in the broad sweep of earth history. Snow and ice are, after all (in a necessary statement of the obvious), simply solid water. As such, they prevail or disappear in response to variations of heat flow within the earth–atmosphere energy system. Change is a constant condition of nature, and nowhere is this more apparent than in the advancing and retreating tides of snow and ice. Man has witnessed these changes through each day, each season, each millennium; for the history of man coincides with one of the rare geological epochs – comprising less than 10% of the earth's history – when snow and ice abound.

Considering that twentieth-century man lives in a relatively active phase of an ice age, it is significant that we know surprisingly little about snow and ice phenomena and the processes associated with them. It is, in fact, only in the last three decades that more than a handful of researchers have focused their attention on solid water phenomena. Nevertheless, we have learned enough to paint a broad, if incomplete, picture of the characteristics, distributions, and processes relating to snow and ice. The following sections briefly cover those subjects, but the treatment is necessarily selective and generalized for reasons of space.

2. Properties of snow and ice

Water is a mineral, and may be defined in terms of its physical characteristics. Most properly, *water* refers to hydrogen dioxide (H_2O) in solid, liquid, or gaseous states. Common usage, however, distinguishes the solid and gaseous forms as ice and water vapour respectively. An unfortunate semantical problem exists when we describe liquid water. This is because we not only use the term water in a general and encompassing sense but also to describe the substance in its liquid state. For the purposes of this article the latter definition will be used; but when quantities of ice or vapour must be described [liquid] water-equivalent units will be used.

A. Physical properties

Crystal form

Ice is characterized by crystals of the hexagonal system, and commonly takes on a variety of prismatic, pyramidal, or dipyramidal forms. Hexagonal symmetry occurs, and is especially prominent in the aggregates of ice crystals which form snowflakes. Crystal size, form, and aggregate structure are highly variable and depend greatly on mode of formation and local environment. Crystals in a glacier, for example, may vary from less than 1 mm to over 1 m in length.

Density

Density relationships between ice and liquid water are anomalous; only a few other substances experience expansion during crystallization. Distilled water has a density of 1·0 at 4° C and 760 mm pressure, which is the standard base for density measurements. At 0° C ordinary ice has a density of 0·92. Depending upon the presence of impurities or gas bubbles and the structural organization of crystals, ice densities vary considerably in nature. Glacier ice, for example, has an average density of 0·84–0·85, while maturely developed lake ice has a density of 0·89–0·90. Sea ice, because of the presence of impurities, has a density of 0·91–0·93. Snow densities are especially variable and dependent on degree of compaction. Density extremes for newly fallen snow range from 0·004 to 0·33, although most meteorological services accept and record an average density of 0·10.

The significance of water's anomolous density behaviour is obvious. Were densities to increase inversely with temperature, ice would sink to the bottom of lake and ocean basins. Since very little of this ice would melt seasonally, there would be a progressive accumulation of ice from depths upward in the earth's lake and ocean basins.

Energy relationships

Energy and water are inseparable and interacting elements of the environment, and changes of state between water and ice account for major energy fluxes. Ice melts to water at 0° C; sea ice at −2° C. Melting ice absorbs 80 g calories of energy for every gram of water melted, an equal amount being released when the process is reversed. The specific heat of ice, depending on temperature, is approximately 0·50, or about half the specific heat of water.

Other properties

Ice is colourless to white to pale blue, and varies from transparent to translucent. Hardness varies with temperature. At −5° C and −44° C, for example, ice has a hardness of 1·5 and 4·0 respectively.

B. Types of snow and ice

The variety of forms taken by snow and ice are too numerous and specialized of definition to enumerate here. It can be generally stated, however, that there are three principal zones of occurrence of snow and ice in the hydrological cycle, each characterized by familiar snow or ice types:

1. Snow or ice which occurs and originates directly in the atmosphere from sublimation or the freezing of previously condensed droplets

Ice clouds and precipitation types such as snow, sleet, and hail are examples. Since these crystals are either precipitated to earth or return to vapour, their existence in the atmosphere is relatively brief.

2. Snow or ice which occurs on the earth's surface

Examples exist at all scales. Frost, rime ice, and glaze ice occur when atmospheric moisture crystallizes directly on the earth's surface. Sea, lake, and river ice result from the freezing of water already present at the surface, while snow cover, névé, firn, and glacier ice result primarily from the accumulation and subsequent metamorphosis of precipitation.

3. Ice which occurs beneath the earth's surface

Most familiar, perhaps, is interstitial ice in soil or detritus. Such ice may be ephemeral, seasonal, or – as in the case of permafrost – relatively permanent.

3. Distribution of snow and ice

In recent geological time – some 10,000–20,000 years ago – continental glaciers covered about 32% of the earth's land surface. Much of that ancient ice has receded and disappeared, and man pretends that he lives in a warm, non-glacial age. Yet the evidence is startlingly contradictory (fig. 8.I.1 and 8.I.2). Some 10% of the land surface remains covered by glaciers, while 7% of the ocean surface is coated by pack or sea ice at their maximum winter extent. An additional 22% of the earth's land surface is underlain by continuous or discontinuous zones of permanently frozen ground. Seasonal snow coats the continents throughout the mid-latitudes; and even at the Equator, the higher mountain summits are capped by snow and ice. The glacial age is not over; it has only diminished in intensity.

Ice and snow remain perennial features of landscape wherever conditions favour survival. The polar latitudes account for most permafrost ice, pack ice, and glacier cover. For example, over 99% of glacier ice is found in Antarctica, Greenland, and islands of the Arctic Archipelago. But alpine regions favoured by high winter precipitation and short, cool summers also support tens of thousands of glaciers and perennial snowfields. Outstanding in this respect are the monsoon-affected Himalayas and, standing in the path of marine air mass flow, the glacierized mountain belts of north-western North America, western South America, and Scandinavia.

4. Snow and ice as an input–output system

Snow and ice which occur at the earth's surface may be schematically viewed as a simple input–output subsystem within the hydrological cycle; that is, the

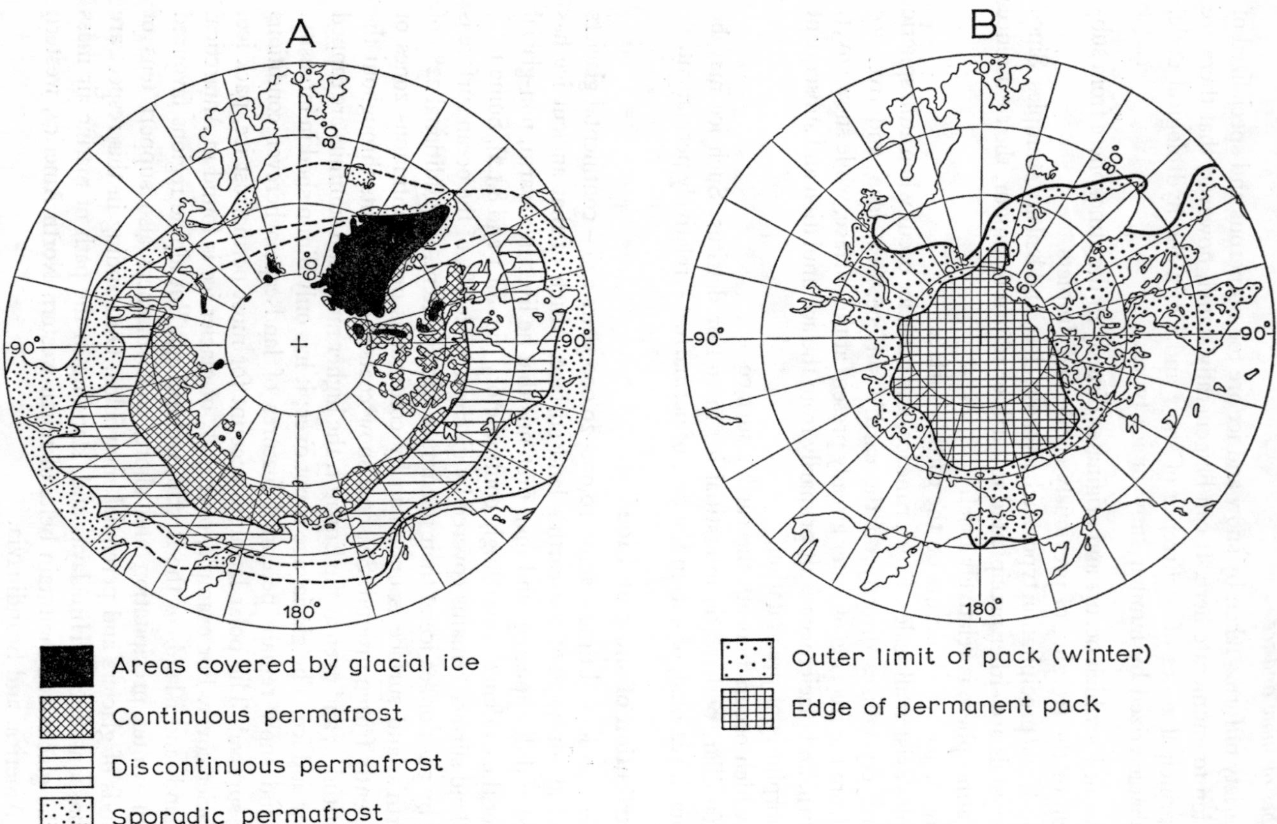

Fig. 8.1.1 Northern hemisphere extent of present glacial ice and permafrost (A), and of pack ice (B) (Partly after Black, R.F., *Bulletin of the Geological Society of America*, Vol. 65, 1954).

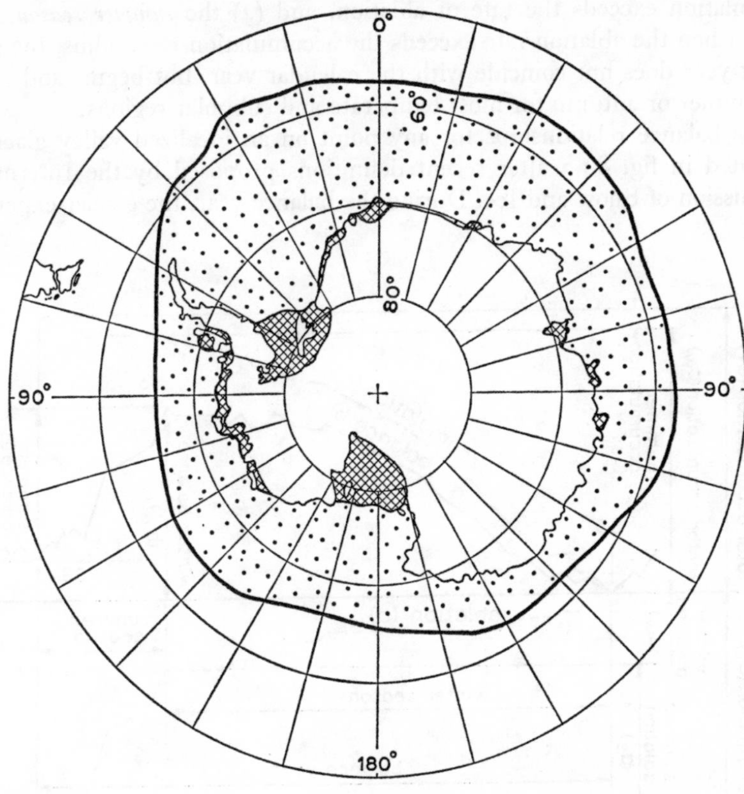

Fig. 8.1.2 Southern hemisphere extent of present ice shelve and pack ice (the central white area indicates the Antarctic ice cap).

growth or diminution of ice and snow cover is a response to net differences between *accumulation* (water input) and *ablation* (water output). Thus, snow or ice covers are open-ended systems whose life spans are dependent upon interacting heat and moisture fluxes across the system interfaces. Glaciers and seasonal snow cover are two widespread and significant examples of this principle.

A. Glacier mass balance

The *mass balance* or *hydrological budget* of a glacier is the net quantity of water gain or loss occurring in a glacier over time – usually a period of one or more glacier balance years. A balance year is the time interval between the formation of two consecutive summer surfaces, where *summer surface* is defined as the time when minimum mass occurs at the site. Mass balance terms vary with time and can be defined seasonally: (1) the *winter season*, which begins when the rate of accumulation exceeds the rate of ablation, and (2) the *summer season*, which begins when the ablation rate exceeds the accumulation rate. Thus, the glacier budget year does not coincide with the calendar year, but begins and ends in late summer or autumn for most temperate and subpolar regions.

Mass balance relationships for any point on an idealized valley glacier are illustrated in fig. 8.1.3 after recent definitions proposed by the International Commission of Snow and Ice. During the balance year, the glacier experiences

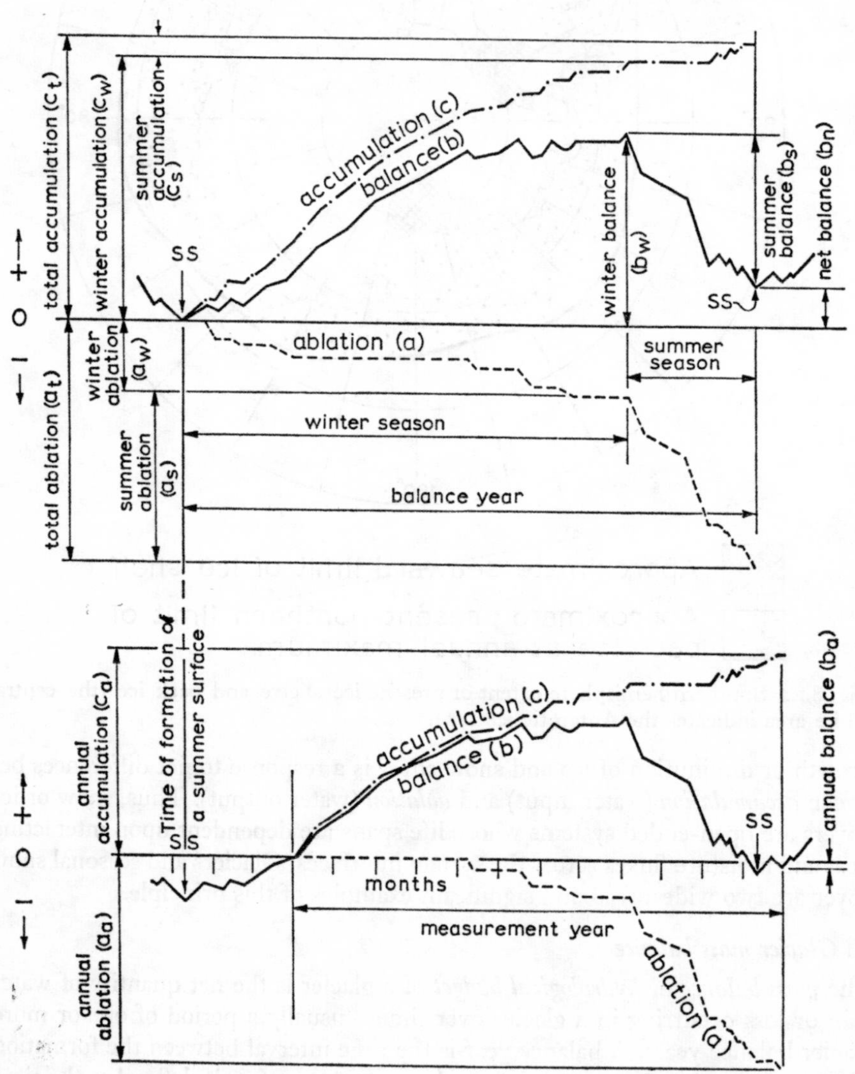

Fig. 8.1.3 Mass balance terms as measured at a point on a glacier (From International Commission of Snow and Ice, 1969).

cumulative accumulation c and ablation a. The balance b throughout the budget year is shown by the solid line and given as $b = c + a$ (where c is always positive and a is always negative). Using additional values from fig. 8.1.3, *net balance* is given as $b_n = c_t + a_t = b_w + b_s$. In this example, since water input is greater than water loss ($c_t > a_t$), the sample point experienced a positive balance year. An equilibrium condition would exist if $c_t = a_t$; a deficit budget if $c_t < a_t$. Aerial mass balance quantities can be similarly calculated by integrating point values over the entire glacier surface.

In fig. 8.1.4 the total glacier is shown at the end of the budget year. Net water gain is included in the snow-covered accumulation zone, and net deficit is encompassed by dashed lines in the ablation zone. If the glacier is in equilibrium the water equivalents of ice flowing across the cross-section under firn line m will

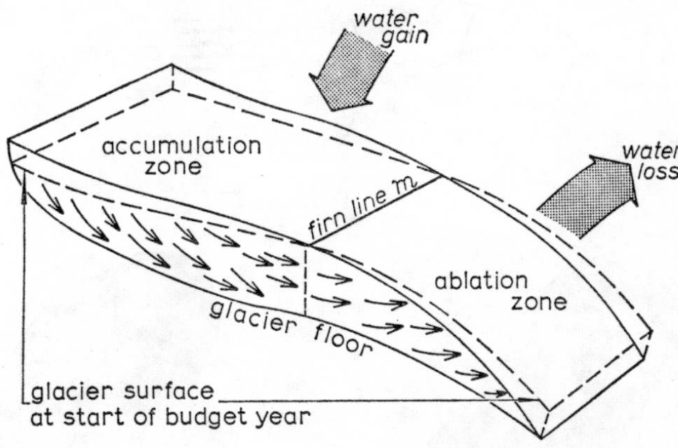

Fig. 8.1.4 Schematic changes in the geometry of a glacier during an equilibrium budget year.

equal ablation and accumulation. Thus, the glacier will retain its original shape and volume. For positive or negative budgets a resultant adjustment of size, shape, and firn line position would occur.

Glacier mass budget can be determined by a number of methods, but the most frequently used technique involves sampling ablation and accumulation over the glacier surface. Figure 8.1.5 illustrates a sampling pattern used on Place Glacier, British Columbia, by Canadian glaciologists. Ablation and/or accumulation is measured at stake positions during the ablation season. Additional soundings of snowpack depth are also accomplished along designated profiles, and variations in density are determined from samples taken in the pits. The data is then integrated to give water-equivalent values of the accumulation, ablation, and budget terms. It is important that field work continue to the end of the glacier year if net budgets are to be calculated. In some cases records of stream discharge below the glacier snout are maintained to check ablation calculations. Such discharge observations must, of course, be corrected for rainfall and snow melt in the catchment area above the glacier.

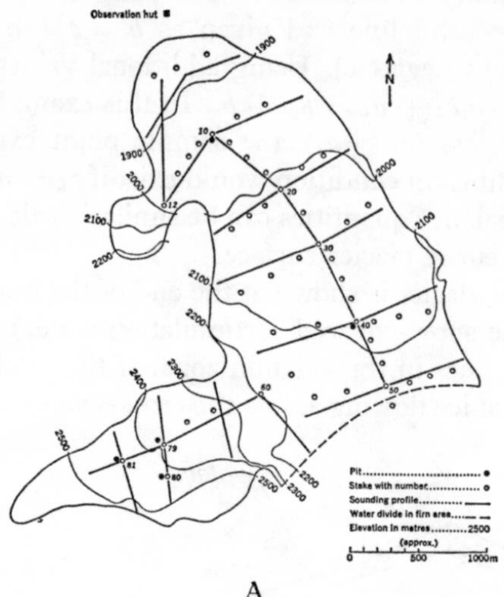

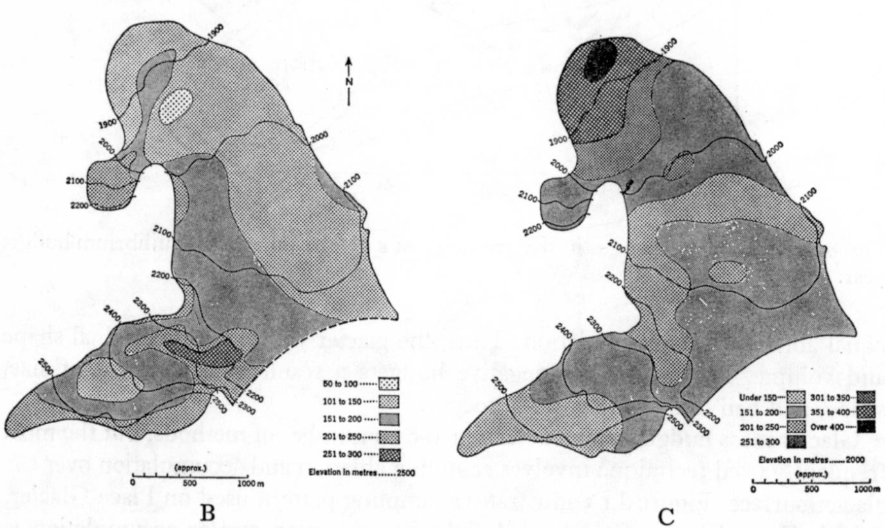

Fig. 8.1.5 Observations on the Place River Glacier, British Columbia, 1964–5 (From Østrem. G., 1966, Mass balance studies on glaciers in western Canada; *Geographical Bulletin*, Vol. 8 (1), pp. 81–112).

A. Location of stakes, pits, and sounding profiles 1964–5.
B. Accumulation map 1964–5.
C. Ablation map 1965, based on readings from 52 stakes.

In summary, glaciers may be viewed as an open-ended and relatively long-termed storage element in the hydrological cycle. Minor fluctuations of glacier systems occur continually, but larger changes are cumulative over decades and centuries. Climatic change is primarily responsible for these mass budget variations, but the interrelationships between climate and glaciers is exceedingly complex and poorly understood. We know that a glacier may grow or diminish in response to changing energy and moisture fluxes across its surface, but we still have difficulty factoring out the relative importance of individual climatic factors. Changes in planetary energy, moisture, and momentum patterns undoubtedly influence mass budgets, but it is also true that local environmental conditions – topography, land–water relationships, landform orientation, etc. – cause microscale and mesoscale variations in climates. Thus, the relative significance of climatic factors such as radiation, wind, cloud cover, and precipitation may differ appreciably from glacier to glacier.

B. Seasonal snow cover

In principle, the seasonal snow cover responds to energy and moisture fluxes as do glaciers. Inputs and outputs determine the mass budget, and the only distinctions are in scale and time. Unfortunately, we can only guess at the amounts of water which are exchanged as the seasonal snow line advances and retreats; no precise broad-scale or global measurements have been made. The seasonal snow cover has, however, a significant impact on human activities. Agriculture, transportation, and flood control are only a few of the activities which are influenced by the size and duration of the snow cover. This information is, in fact, of such importance that we may expect in the near future – through use of satellite sensors – daily reports on the budget, condition, and migration of seasonal snow.

REFERENCES

DYSON, J. L. [1962], *The World of Ice*; (Alfred Knopf, New York), 292 p. (An interesting and superbly illustrated book treating all aspects of snow and ice.)

FRASER, C. [1966], *The Avalanche Enigma*; (John Murray, London), 301 p. (A fine general source on snow characteristics, avalanche mechanics, and the impact of avalanches on human activity.)

INTERNATIONAL COMMISSION OF SNOW AND ICE [1969], Mass balance terms; *Journal of Glaciology*.

SHARP, R. P. [1960], *Glaciers* (The University of Oregon Press, Eugene, Oregon), 78 p. (An excellent account of glacier structure, flow, and mass budget, made easily understandable for the non-specialist.

WEEKS, W. and ASSUR, A. [1967], *The Mechanical Properties of Sea Ice*; Cold Regions Scientific and Engineering Series, Part II-C3, U.S. Army Cold Regions Research and Engineering Lab., (Hanover, New Hampshire), 80 p. (An excellent technical summary of lake and sea ice properties.)

9.I. The Flood Hydrograph

JOHN C. RODDA
Institute of Hydrology, Wallingford

I. The hydrograph

Where the quantity of runoff from a river basin is measured continuously, a graph of flow as a function of time is obtained. This discharge hydrograph expresses the sequence of relationships that occur between runoff and the other components of the basin water balance, together with their adjustments to the physical characteristics of the basin.

During dry periods the flow of a river decreases exponentially. This base flow

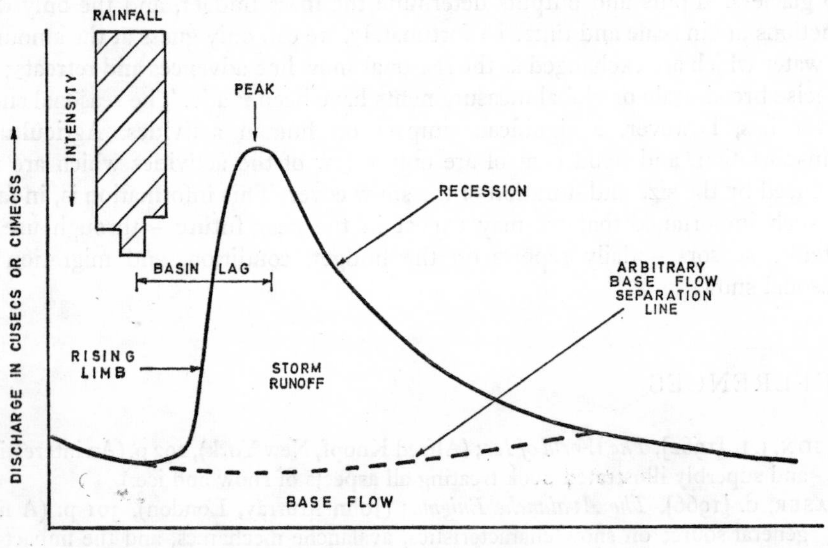

Fig. 9.I.I Components of a hydrograph.

(fig. 9.I.I) continues until rain falls, when the rate of discharge increases rapidly to a peak or crest. This point is arrived at when the quantity of water draining to the gauging station from the basin has reached a maximum. The peak is usually attained shortly after rain has ceased, and thereafter flow is largely determined by the amount of water in storage. This stored water will be held in the soil and in the bedrock, and will have resulted from recent infiltration and percolation from earlier rain. The rate of withdrawal from storage controls the shape of the recession limb of the hydrograph, this relationship between storage and discharge being expressed as:

$$Q_t = Q_0 K^t \qquad (1)$$

where Q_t is the discharge at time t after some instant when the discharge was Q_0, K being a recession 'constant', which is less than unity. Further rain during recession can cause two or more peaks, but without it, discharge decreases until the extra water in storage due to the recent rain has been depleted and the flow is approaching its original volume.

Thus the discharge hydrograph consists of a series of irregular saw-tooth-shaped fluctuations superimposed on a gently undulating section. These two components are usually defined as storm runoff (or the flood hydrograph) and base flow, the latter being attributed to ground-water discharge. A further analysis of the hydrograph has been made by separating storm runoff according to the path to the river. The initial rise in the hydrograph is attributed to channel precipitation – water falling directly on to the connected water surfaces of a basin– while the bulk of the increase in discharge is said to be caused by surface runoff. Interflow is the third component, and this water, which moves laterally through the upper soil horizons until it is intercepted by a stream channel, is particularly important during recession.

These concepts may be useful for descriptive purposes, but their scientific basis is limited. For example, surface runoff can only occur when the rainfall intensity exceeds the infiltration rate, yet in many storms this never happens, no surface runoff results but the hydrograph rises sharply. Indeed, for most hydrographs it is difficult to distinguish between storm runoff and base flow, let alone the other components, so most modern approaches concentrate on this two-fold division, employing it in unit hydrograph analyses.

2. Hydrograph shape and magnitude

The shape and dimensions of the hydrograph are controlled by a variety of factors, many being interrelated. These factors can be divided into two main groups; those of a permanent nature and those that can be classed as transient (fig. 9.1.2). The first group in general represent the characteristics of the basin, while the second is associated with climate and related features. Of course, one or two factors fall into either category, and there are others that may change from one group to the other – usually as a result of man's activities.

While the factors that influence the characteristics of the hydrograph are for the most part readily recognizable, relatively few studies have aimed at establishing quantitative relations between the flood and these factors. Yet a precise method of predicting the size and shape of the flood hydrograph would be invaluable to the hydrologist, in place of the somewhat dubious means that he is forced to employ at times. Such a method could be utilized for flood-warning purposes and to provide the basic information for the design of spillways, culverts, bridges, and similar structures. For flood-warning purposes, however, it would be necessary to know *when* a flood is likely, as opposed to the

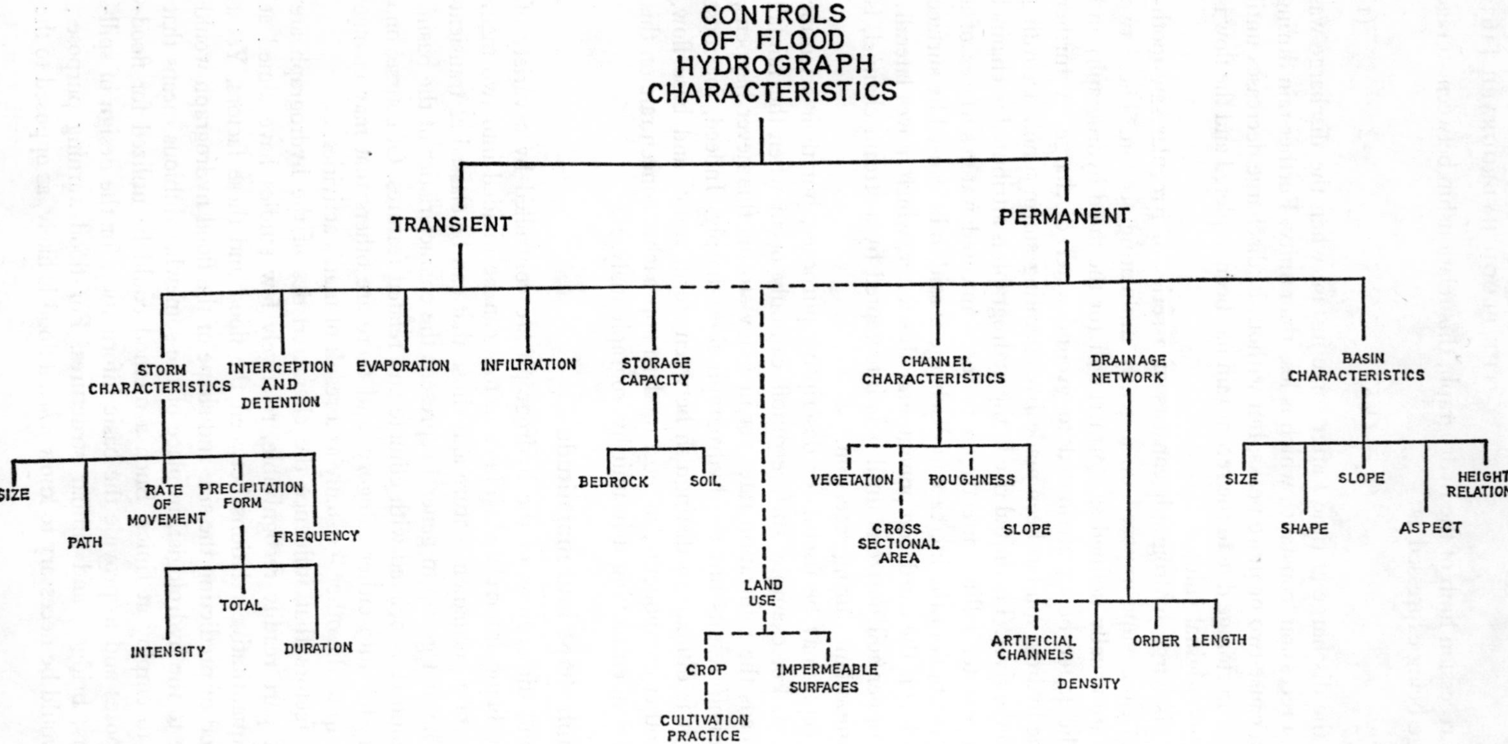

Fig. 9.1.2 Controls of flood hydrograph characteristics.

design need, which requires information on *how often* a flood of a particular magnitude would occur.

Some floods are caused by dam bursts, earth movements, and high tides, but these are rare by comparison with the floods due to intense rain or rapid snow melt. In the case of rain-induced floods the path of the storm in relation to the alignment of the basin, the size of the storm, and its rate of movement are important, as well as the intensity and amount of rain. Once the rain has reached the surface of the basin, the rate and amount of runoff will be influenced by such factors as the current evaporation and infiltration rates, the soil moisture status, and the type of land use. Movement of this water into rills and brooks and the passage of the flood wave to the gauging stations are governed by a series of factors, from the basin morphometry to the hydraulic and biological characteristics of the channel system. These factors control not only the form of the hydrograph but also the time interval between the rain and the flood. Basin lag and time of concentration are indices of these time-response characteristics of a basin; the first being the time between the centre of mass of rain and centre of mass of runoff, and the second, the time taken for water to reach the gauging station from the most distant point in the basin. However, even though these features are important, far more attention has been given to estimating the magnitude of the flood peak.

3. Factors influencing the magnitude of the flood peak

The interdependence of factors, the difficulties of quantifying and identifying them correctly, and the problems of establishing meaningful statistical relationships between variables have to be overcome in determining the controls of the flood peak. Interdependence is probably the major obstacle, because virtually all the controls are related in some way to one another. Hence the usual approach is to select as variables factors that on physical grounds are least likely to be interdependent. There are difficulties in quantifying some of the factors, such as vegetation or land use, while in many cases measurements of others are simply not available (e.g. infiltration rates and rainfall intensities). Often the only material that can be obtained, other than river-flow records, is what can be derived from map analyses or from aerial photographs. This emphasizes the importance of quantitative geomorphology to this branch of hydrology; although few of the morphometric properties that can be assessed have been shown by objective methods to be important controls. Basin size, basin and channel slope, and various measures of the channel system can be determined from maps. These and similar factors have been demonstrated to be significant and have been incorporated into a range of 'flood formulae' by means of standard statistical techniques producing expressions of the following type:

$$Q_t = aX_1^b\, X_2^c\, X_3^d \qquad (2)$$

where Q_t = the T-year annual peak discharge;
$a \dots d$ = regression coefficients;
X_1 etc = the factors controlling the flood peak.

A. Basin area

Other factors being equal, the larger the size of the basin, the greater the amount of rain it intercepts and the higher the peak discharge that results. This rather obvious conclusion has been the basis for a large number of flood formulae in the general form:

$$Q = CA^n \tag{3}$$

where Q = peak discharge;

A = basin area;

C = a constant that varies according to the land use or topography of the basin;

n = a constant that has a range from 0·2 to 0·9, depending on climate to some extent.

One of the many examples of this type of approach is a study of the mean annual flood (Q_m) in a number of basins in England and Wales, where the area-discharge relationship was determined as:

$$Q_m = CA^{0 \cdot 85} \tag{4}$$

However, there are a number of other factors that are partly dependent on basin area and must be in some way accounted for by the constants in this type of relationship. For example, larger basins are usually less steep than smaller ones, and this applies to other factors, such as channel slope and rainfall intensity. As a consequence, when these additional characteristics of the basin are employed the inclusion of basin area with them usually gives the most meaningful result.

B. Basin shape

Basin shape is of obvious importance in influencing peak flow and other hydrograph characteristics, although it is a feature which is difficult to express numerically. However, a number of shape indices have been developed, some of the best known being the form factor and Miller's circularity ratio. The former is the ratio of average width to axial length of the basin, while the latter demonstrates the circularity of a basin by:

$$R_c = \frac{A_b}{A_c} \tag{5}$$

where A_b area of basin;

A_c area of the circle having the same length of perimeter as the basin.

This expression has a value of unity for a circular basin, while for two basins of the same size the flood potential would be considered greatest for the one with the smallest circularity coefficient. This index was employed in analysing the flow from a number of Appalachian basins, but it was found to have a low correlation with peak discharge. On the other hand, in the same study, peak discharge was found to be highly correlated with the longest length of basin as measured from the head of the basin to the stream gauging station. This and other indices of shape have been criticized on the grounds that they do not approach the ideal pear-shaped basin – the lemniscate being put forward as offering a better comparison.

C. Basin elevation

The altitudinal extent of the basin above the gauging station exercises direct and indirect control over the magnitude of the flood peak. With basin slope and several additional factors, it determines the proportion of runoff, and indirectly it influences a number of other important controls, such as precipitation, temperature, vegetation, and soil type. However, it is difficult to compute a single term which gives a meaningful measure of basin elevation. Indeed, several studies have shown the various indices that have been devised to have no significant relation to the size of the flood peak.

D. Basin slope

Slope, like elevation, is an obvious control of peak discharge, but again it is a factor which is difficult to interpret meaningfully. Some methods of slope assessment are extremely involved and require measurement of length of all contours in a basin, or counting the number of intersections between contours and a grid overlay. Others are relatively simple, but the importance of these basin slope indices has been difficult to establish, whereas measurements of channel slope have been proved significant. One slope index (S) devised for a study of fifty-seven British drainage basins showed little significance, even when log S was correlated with log Q_m. However, when slope was combined with basin area, as below, a coefficient of multiple correlation of +0·93 resulted:

$$Q_m = 0{\cdot}074A^{0{\cdot}74}S \qquad (6)$$

E. The drainage network

The several characteristics of the flood hydrograph hinge on the efficiency of a basin's drainage system. A quick rise to a high peak is the mark of a well-developed network of short steep streams. Conversely, a minimal response to intense rain usually reflects an incipient channel system. How a particular basin relates to these extremes of development can be assessed in terms of linear aspects of the drainage network, the areal relationships of the system, and the various channel gradients.

Linear aspects of the channel system are expressed in terms of stream order, bifurcation ratio, stream length, and length of overland flow. Other than longest length of stream channel, none of these measures, by themselves, have been shown to exercise extensive control over the flood peak. On the other hand, their inclusion with other factors has reduced the error of estimate of peak flow, and this also applies to areal relationships and channel gradients. For a study of New England floods, ninety-three slope factors were computed, main-channel slope being found the most significant variable. In the same study peak flow showed no relation to drainage density, once channel slope and storage area had been taken into account. On the other hand, the drainage density term improved the prediction equation in a flood study for the United Kingdom, and it has been shown to correlate with the mean annual flood in other studies.

Apart from channel slope, no other measures of the hydraulic nature of the channel have been included in flood peak studies, although roughness and wetted perimeter would be important terms. Similarly, there is an absence of expressions for the character of the in-channel and riverine vegetation, features that can also influence the nature of the hydrograph.

F. Climatic factors

In latitudes where snow melt combines with rain to produce a spring flood maximum, snow depth and temperature are important controls of peak flow. Elsewhere, the magnitude of the flood is related to the rainfall that provokes it, coupled with the current storage capacity of the basin. There are difficulties in expressing this rainfall, because each storm is typified by a differing set of magnitude, duration, and intensity relationships, as well as those of frequency, distribution, and areal extent. One way of avoiding these difficulties is to employ the basin's mean annual rainfall as an index of its flood susceptibility. This was the basis for a second part to the study of floods in Britain, where the slope factor (S) was replaced by mean annual rainfall (R) to give:

$$Q_m = 0{\cdot}009A^{0{\cdot}85}R^{2{\cdot}2} \tag{7}$$

However, this term is not a sufficiently realistic measure of the flood-producing rain, and several more likely ones have been determined. These include various intensity–duration parameters and one of rainfall frequency. For example, in the New England study of floods, referred to previously, the intensity index showing the highest correlation with peak discharge proved to be the daily maximum rainfall of the same frequency as the flood. Further improvements could be made by the inclusion of factors indicative of antecedent conditions or the availability of storage, but such terms have rarely been employed.

G. Vegetation and land use

Speculation about the effects on runoff of felling a forest, or one of the other land use changes brought about by man's activities, has continued for many centuries. Only at the end of the nineteenth century were experiments commenced to determine what hydrological differences resulted from alterations in land use, but since then over a hundred of these experiments must have been conducted in various parts of the world. In fact, none of the other controls of runoff have received comparable attention. The classical approach is to alter the land use on one of a pair of otherwise identical basins after an initial calibration period, then to ascribe differences in runoff patterns to the contrasting land usage. The basin maintained in its original state acts as a control, so that extraneous influences, such as climatic change, can be identified. This is the disadvantage of the single-basin approach, because the land use change that is made after the calibration period can be confused with trends in rainfall or the long-period variations in another element of climate. Results from these experiments are often not representative of larger areas, and few, if any, vegetation factors have been employed as variables in assessing peak discharge. Nevertheless, the information gained from these studies, which are as yet far from complete, is of considerable interest and importance (Penman, 1963; Sopper and Lull, 1967) from the floods point of view and in terms of water resources.

H. Comprehensive formulae for estimating the flood peak

Various measures of basin morphometry and climate have been combined in formulae for assessing the T-year peak discharge. All such formulae include a basin area factor, and most contain some index of rainfall intensity and frequency, in addition to differing measures of several morphometric characteristics. The efficacy of each extra term is demonstrated by a reduction in the standard error of estimate and a rise in the coefficient of multiple correlation in the examples shown below.

Author	*No. of basins*	*Location*	*Formula*
Potter	51	Allegheny–Cumberland Plateau	$\log Q_{10}{}^1 = -1{\cdot}4 + 0{\cdot}17 \log A^1 - 0{\cdot}55 \log T + 0{\cdot}93 \log P + 0{\cdot}45 \log S$
Morisawa	15	Appalachian Plateau	$\log Q_{10}{}^1 = -8{\cdot}96 + 0{\cdot}54 \log A^1 + 0{\cdot}72 \log t + 4{\cdot}24 \log P - 0{\cdot}29 \log S$
Benson	164	New England	$\log Q_{2{\cdot}33} = 0{\cdot}4 + 1{\cdot}0 \log A + 0{\cdot}3 \log Sl - 0{\cdot}3 \log St + 0{\cdot}4 \log F + 0{\cdot}8 \log O$
Rodda	26	United Kingdom	$\log Q_{2{\cdot}33} = 1{\cdot}08 + 0{\cdot}77 \log A + 2{\cdot}92 \log R_{2{\cdot}33} + 0{\cdot}81 \log D$

$Q_{10}{}^1$ = peak discharge (cusec/acre) for a ten-year recurrence interval.
$Q_{2{\cdot}33}$ = peak discharge (cusec) in the mean annual flood.
A^1 = basin area (acres).
P = rainfall intensity factor.
T = topography factor.
S = rainfall frequency factor.
t = topography factor combining measures of relief, circularity, and first-order stream frequency.
A = basin area (square miles).
Sl = main channel slope (ft/mile).
St = % of surface storage area plus 0·5%.
F = average January degrees below freezing (° F)
O = orographic factor.
$R_{2{\cdot}33}$ = mean annual daily maximum rainfall (in.).
D = drainage density (miles/square mile).

Even the inclusion of three or more independent variables covering a wide range of climate and basin characteristics leaves considerable differences between observed and predicted peak discharges. Indeed, it has been suggested that one year of discharge records is more valuable for predicting the T-year flood than innumerable estimates of flood magnitude–frequency relations. It could be that the problem of prediction will never be solved by application of the basic physical laws, even where these are fully understood, because the physical system is far too complex. Some hope of success is offered by simplification of the system and its input through the use of quantitative models to simulate the major processes and interactions within the basin.

4. Model simulation of basin discharge characteristics

The use of models in the study of various aspects of the hydrograph has been common for some time. The model system consists of a limited number of parts, analogous to the most important features of the prototype, and it operates by transforming numerical input data into a quantitative representation of the behaviour of the prototype. Parameters in the model, which are obtained initially by comparison with the prototype, are adjusted so that the two outputs match. This optimizing technique is not a simple procedure, because a number of parameters have to be adjusted, and it is not completely clear which mathematical methods of optimization are best. However, if all basins could be described fully by an expression of the type

$$Q = PX_1{}^q \tag{8}$$

in which P is the percentage run-off and q is a rainfall intensity factor, then for one particular basin the rainfall and discharge records would be scrutinized and the optimum values for the parameters P and q sought by a sequence of adjustments, to improve the agreement between calculated and observed peak discharges – the agreement being measured objectively. This two-parameter model is obviously far too simple a representation of a basin, and it ignores many of the physical processes involved in runoff. Yet the parametric approach has the advantage of not requiring detailed measurements of the entire range of processes involved, because where a significant physical effect is neglected, the model will show this by its failure. The use of digital computers in these simulation studies permits quite complex models to be employed, but there are other examples where parametric methods are employed, such as co-axial correlation and the unit hydrograph.

A. Co-axial correlation

The co-axial method of graphical correlation has been widely applied for the prediction of the total volume of storm runoff. A range of factors have been incorporated in the diagrams and in the example shown (fig. 9.1.3). A four-parameter model was adopted for computing the storm run-off in the River Thames at Teddington.

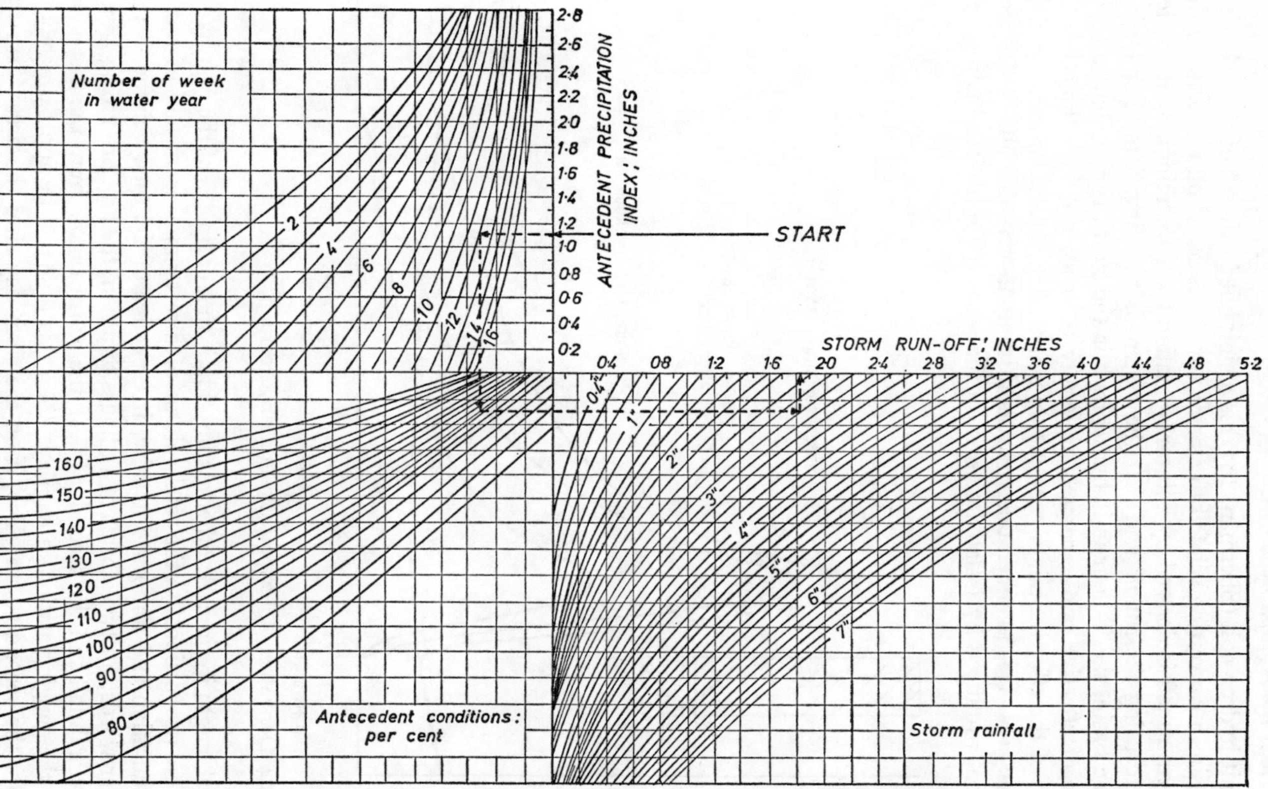

Fig. 9.1.3 Co-axial diagram for computing storm runoff for the River Thames at Teddington (After Andrews, 1962).

B. The unit hydrograph hypothesis

The unit hydrograph (Sherman, 1932) is a simplified concept of the behaviour of a basin in converting rainfall to stream flow. It is based on the premise that the storm runoff is derived by a linear operation from the rainfall that is effective in causing the runoff and that the system is time invariant. For a given drainage basin, the T-hour unit hydrograph is the storm runoff due to a unit volume of effective rain generated uniformly in space and in a time T – the volume of the effective rainfall commonly being taken as 1 in. or 1 cm over the drainage area. It is assumed that the runoff from effective rainfalls of the same duration, produced by isolated storms on the same basin, causes hydrographs

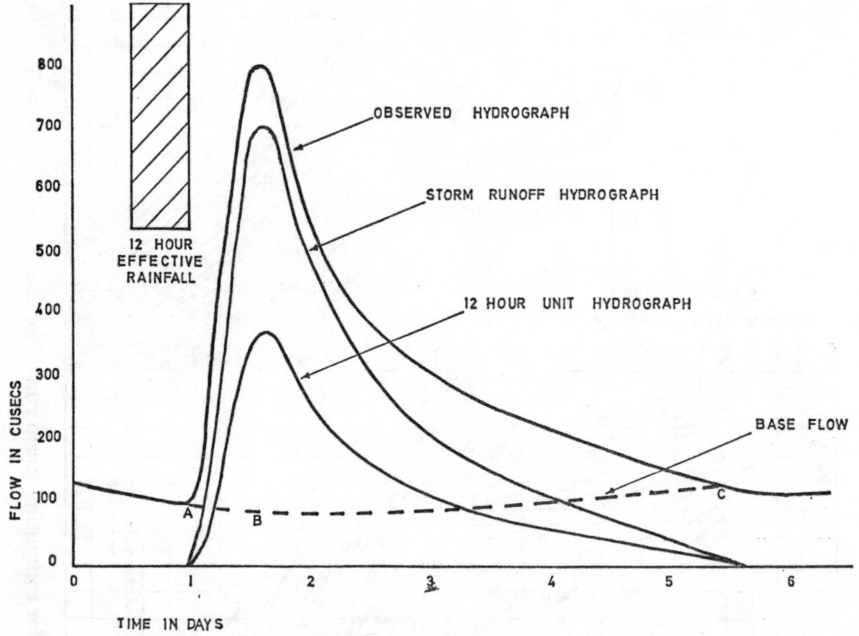

Fig. 9.1.4 Derivation of a hypothetical unit hydrograph.

of equal length in time. Another assumption is that ordinates of the unit hydrograph are proportional to the total volume of direct runoff from falls of rain of equal duration and uniform intensity, irrespective of the total volume of rain. In other words, for two storms of 15 hours duration, the first producing 2 in. of effective rainfall and the second 3 in., if the storm runoff from the first storm passes the gauging station in ten days, then this time will be approximately the same for the second storm. Also, if on the sixth day after the first storm 15% of the total storm runoff passes the gauging station, then this will apply to the second storm. It is evident that for natural basins these assumptions about the unit hydrograph cannot be justified completely, but for many purposes the results produced by the theory are acceptable.

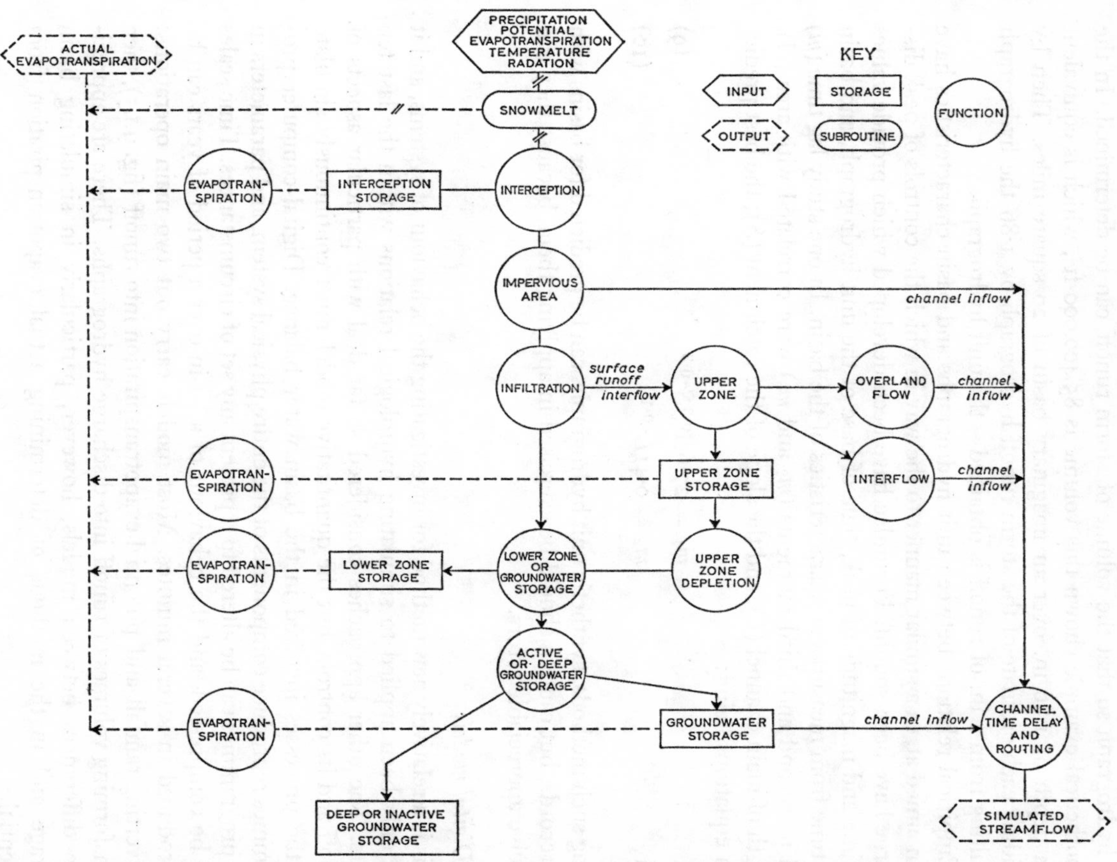

Fig. 9.1.5 Stanford Watershed Model IV flowchart (After Crawford and Linsley, 1966).

To derive a unit hydrograph for a particular basin, records of rainfall and discharge are examined for an isolated storm with reasonably uniform rainfall. For such a storm (fig. 9.I.4) the base flow (*ABC*) is separated from the remainder of the hydrograph so that the volume of storm runoff can be determined. In the hypothetical example shown this volume is 85,400,000 ft^3, which is equivalent to a depth of 1·86 in. over an imaginary basin of 20 square miles. Then by dividing each ordinate of the storm runoff hydrograph by 1·86 the hydrograph resulting from 1 in. of runoff is obtained – the unit hydrograph.

Empirical relations between unit hydrographs and basin characteristics have been aimed at, in a similar manner to the way in which the controls of peak discharge have been sought. Formulae have been developed which provide values of time and magnitude of peak, time of base of the unit hydrograph, and basin lag time from measures of characteristics of the basin. In one study lag time (m) and its second and third moments (m_2 and m_3) were correlated with area (A), length of main channel (L), and the slope of the catchment (S); the best prediction equations being:

$$m_1 = 27{\cdot}6A^{0{\cdot}3}S^{-0{\cdot}3} \tag{9}$$

and

$$m_2 = 0{\cdot}41L^{-0{\cdot}1} \tag{10}$$

using such methods, synthetic unit hydrographs can be predicted for basins with no records, but caution should be exercised in applying them, because they can involve appreciable errors.

C. Digital models

This is a relatively new method for investigating the behaviour of systems, and it has only been applied to simulating hydrological relations within the last ten years. The other approaches considered so far deal with particular aspects of runoff and its controls, but the quantitative model must continuously simulate all the processes involved in the basin water balance. Digital-computer programmes reproduce components of the entire physical system, and parameters in the programmes can be altered to represent any set of circumstances. Time scales can be compressed, and the behaviour of a basin over a period of years can be reproduced in several minutes. Most models carry out two main operations: converting rainfall and potential evapotranspiration into runoff (fig. 9.I.5), then transforming volumes of runoff into discharge hydrographs. There are appreciable differences between models, however, particularly in simulating basin storage and in the methods of determining actual evapotranspiration from potential.

This is a very rapidly developing field, where a considerable effort is being made. Like all other approaches that aim at an understanding of the controls of the flood hydrograph, its limitations lie mainly in the veracity of the basic data.

REFERENCES

ANDREWS, F. M. [1962], Some aspects of the hydrology of the Thames Basin; *Proceedings of the Institution of Civil Engineers*, **21**, 55–90.

BENSON, M. A. [1962], Factors influencing the occurrence of floods in a humid region of diverse terrain; *U.S. Geological Survey Water Supply Paper 1580-B.*

CARLSTON, C. W. [1963], Drainage density and streamflow; *U.S. Geological Survey Professional Paper 422-C*, 8 p.

COLE, G. [1966], An application of the regional analysis of flood flows; *The Institution of Civil Engineers, Proceedings of the Symposium on River Flood Hydrology* (London), pp. 39–57.

CRAWFORD, N. H. and LINSLEY, R. K. [1966], *Digital Simulation in Hydrology: Stanford Watershed Model IV*; Department of Civil Engineering, Stanford University, Technical Report No. 39, 210 p.

MORISAWA, M. E. [1959], Relation of quantitative geomorphology to stream flow in representative watersheds of the Appalachian Plateau Province; *Office of Naval Research Project NR 389–042, Technical Report 20, Department of Geology, Columbia University, New York*, 94 p.

NASH, J. E. [1960], A unit hydrograph study, with particular reference to British catchments; *Proceedings of the Institution of Civil Engineers*, **17**, 249–82.

NASH, J. E. and SHAW, B. L. [1966], Flood frequency as a function of catchment characteristics; *The Institution of Civil Engineers, Proceedings of the Symposium on River Flood Hydrology* (London), 115–36.

NASH, J. E. [1967], The role of parametric hydrology; *Journal of the Institution of Water Engineers*, **21**, 435–74.

PENMAN, H. L. [1963], *Vegetation and Hydrology*; Commonwealth Bureau of Soils, Harpenden, Technical Communication No. 53, 124 p.

POTTER, W. D. [1953], Rainfall and topographic factors that affect runoff; *Transactions of the American Geophysical Union*, **34**, 67–73.

RODDA, J. C. [1967], The significance of characteristics of basin rainfall and morphometry in a study of floods in the United Kingdom; *UNESCO Symposium on Floods and their Computation* (Leningrad).

SHERMAN, L. K. [1932], Streamflow from rainfall by the unit-graph method; *Engineering News-Record*, **108**, 501–5.

SOPPER, W. E. and LULL, H. W., Editors [1967], *International Symposium on Forest Hydrology: Proceedings of a National Science Foundation Seminar* (Pergamon Press, London), 813 p.

10.1. River Regimes

ROBERT P. BECKINSALE
School of Geography, Oxford University

The regime of a river may be defined as the variations in its discharge. In its widest sense the regime involves all occurrences and is portrayed by a curve based on continuous or hourly observations. Such curves, however, present complicated problems of analysis and for some purposes the discharge variations are better expressed by graphs of mean monthly flow. When used for critical purposes, such as the delimitation of hydrological regions, the ideal seasonal regime hydrograph (station-model) would show additionally, for each month and for the year as a whole:

1. the mean flow;
2. the mean maxima and minima; and
3. the absolute maximum and minimum.

It would also be helpful to insert (not as a curve) the absolute *daily* maximum and minimum recorded during the period.

However, for general comparative purposes and for global classifications the monthly means seem adequate and, indeed, such simple data is still not available for large areas. On this broad scale comparison is facilitated by expressing the mean monthly value either as a ratio of the mean monthly flow for the year (taken as 1) (as in figs. 10.1.5 and 10.1.6), or as a percentage of the mean annual flow (taken as 100) (as in figs. 10.1.2 and 10.1.3). If the actual quantities are stated as one of the ordinates this method does not lose much in practical utility, particularly if the mean annual total flow is also stated somewhere on the regime hydrograph. Only by insisting on actual quantity as well as on comparative ratios will the possibilities of inter-regional water exchanges be kept constantly before civil engineers and planners.

1. Factors controlling river regimes

Seasonal variations in the natural runoff of a drainage basin depend primarily on the relationships between climate, vegetation, soils and rock structure, basin morphometry, and hydraulic geometry. Of these only rock structure and, to a lesser extent, basin size can be strictly independent of climate. It should be stressed that the features of basin morphometry and hydraulic geometry are only of direct relevance to the seasonal regimes of large river basins.

A. Climate

The direct climatic control over river regimes lies in the difference between the monthly sequence of precipitation (positive) and the values of insolation or solar radiation (negative), controlling evaporation. Normally water-vapour pressure is highest and rainfall most abundant in the summer half-year when evaporation and evapotranspiration are also greatest (fig. 10.1.1(*b*)). A few restricted areas, however, have most precipitation in the winter half-year because locally moist maritime airflow, associated with frontal depressions, is strongest then. The direct relation between rainfall and insolation is also departed from in subsident or anticyclonic air masses, and in sunny tropical and sub-tropical regions with prolonged upper-air (*cT*) subsidence (or even with local lee-wave subsidence behind high topographic barriers) potential evaporation often far exceeds precipitation and permanent streams are absent.

In continental areas with snow-cover in the cold season, lower air subsidence creates a shallow anticyclonic air mass (*cP*), beneath which chances of precipitation remain negligible until a warmer air mass intrudes. Thus frigid *cP* air masses, with very low moisture content, have almost the same impediment on precipitation as hot subsident *cT* air masses do, but their low evaporation rate and snow retention ensure that they have eventually a *positive* effect on runoff. Beneath cold *cP* air masses soils usually freeze to a depth of several metres, and when the snow melts in spring runoff may for a while be very great from above the still-frozen sub-soil. In such climates vast areas are underlain by permafrost, above which the soil thaws out in summer. The top of the permafrost layer is usually deeper in river valleys, but in extremely cold northerly areas it remains close to the surface all the year.

The direct effect of climate on a river's surface is also significant, as it includes its freezing as well as the direct channel precipitation it receives and, more important, the evaporation it loses – both of which increase when the channel widens, anastomoses, or winds excessively. These channel-precipitation and evaporation factors vary in potency with the increase in channel area, and from region to region. In humid basins such expansions of channel surface give a net channel *gain* of the water-surface area increase multiplied by the difference between the annual precipitation and the annual floodplain surface runoff (which would have flowed directly into the channel in any event). In arid and semi-arid regions the evaporation increase would probably be equivalent to the increase in water-surface area multiplied by the local evaporation rate, since extra channel length or width would usually be associated with channel shallowing and warming. Such losses can be enormous. Thus on the upper Niger, where the river enters upon and flows north-eastwards over a sedimentary plain (fig. 10.1.2), its channel spreads over a wide area, and evaporation soon lowers its volume from 1,545 m^3/sec at Koulikoro to 1,146 m^3/sec at Mopti about 300 miles downstream.

B. Vegetation

Except in deserts, the influence of climate cannot be divorced from that of vegetation, which must be viewed as thermally driven chains of cells that conduct solutions from the soil to aerial surfaces, and thence to the atmosphere. Hence plant growth normally leads to great losses of water from the soil. The extent to which plants initiate other processes which may offset these transpiration losses is small. The turbulential uplift of a moist airflow athwart the edge of a tall forest may slightly increase rainfall. Similarly, tall plants extract from air masses at or near condensation point (cloud and fog) rain-drip which is of slight significance locally. But as a rule vegetation greatly decreases runoff, and the less the vegetation, the more abundant and rapid the runoff will be. A close plant-cover, at least for a while, modifies the violent effects of heavy rains and nullifies the hydrological effects of very light showers.

Vegetation actually growing in and on river channels may be regarded as a kind of surface roughness which may markedly reduce the capacity of the channel and retard the flow. In shallow streams the growth of hydrophytes raises the surface-level either all the year or in the warm season, and in some large tropical rivers (such as those in the sudd-hindered Bahr-el-Ghazel) a considerable retardation of flow and loss of discharge occur directly due to vegetation.

C. Soils and rock structure

From the point of view of regimes the chief properties of soils are their permeability and their water-holding capacity. These factors work in conjunction with climate, vegetation, and relief to control the amount of sub-surface water that eventually reaches the streams either as springs or as seepages. Hydrographs are often drawn to show the proportion of total runoff that originates as sub-surface flow and ground water. For example, the monthly sequence of the soil–water balance of the River Havel is shown in fig. 10.1.1(*b*).

Some soils and rock structures, unless coated by a continuous clay cover, are highly permeable and have a large water-holding capacity. Among the chief of these are some varieties of porous limestones and of coarse-grained sandstones, and certain well-jointed igneous rocks, notably basalt. Such rock formations, up to a certain limit (which may never be reached locally), tend to even out ground-water discharge, especially during dry seasons. In time of prolonged or heavy rain, after the underground storage capacity becomes almost full or the water-table rises high above spring-heads, this moderating influence often lessens and depends largely on the relative speed, directness, and convergence of underground flow towards the main exits.

D. Basin morphometry and hydraulic geometry

The shape and gradient of landforms greatly influences runoff, which as a rule increases in amount and rapidity with increase of slope. Conversely, flat areas, especially where marshes and lakes occur, tend to accumulate water and to

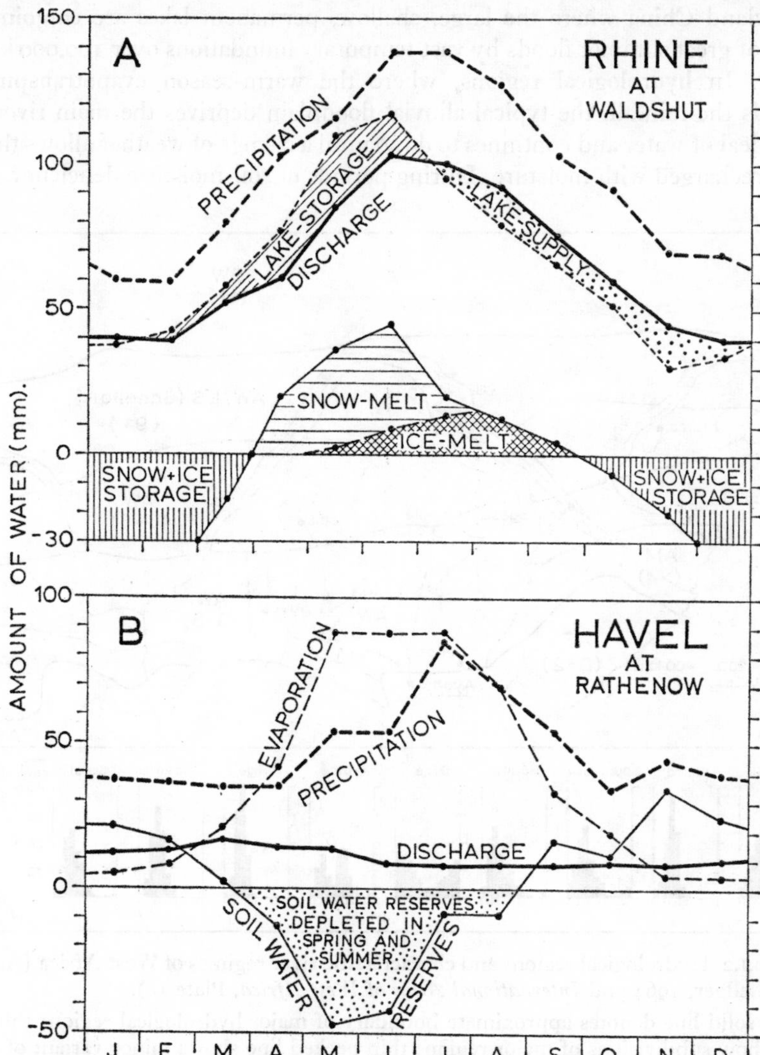

Fig. 10.1.1 Annual water-balance of two rivers.

A. Upper Rhine at Waldshut, showing influence of Alpine snow and ice, and of Lake Constance.

B. River Havel, a tributary of the Elbe, showing relation between precipitation, evaporation, runoff, and soil-water reserves (Adapted from Wundt, 1953).

modulate the regime downstream. The regulating effect of lakes is abundantly documented. Figure 10.1.1(*a*) indicates the moderating influence of Lake Constance on the flow of the Rhine downstream at Waldshut.

The influence of alluvial floodplains resembles that of lakes only during floods, when enormous quantities of water are absorbed and stored above and below ground. The classic example is the lower Yangtze-Kiang and the Hwai Ho in lowland China where the large, shallow, permanent lakes are conjoined in times of great summer floods by vast temporary inundations over 100,000 km^2 in extent. In hydrological regions, where the warm-season evapotranspiration exceeds the rainfall, the typical alluvial floodplain deprives the main river of a great deal of water and continues to do so until a change of weather allows the soil to be recharged with moisture. During periods of soil-moisture deficiency many

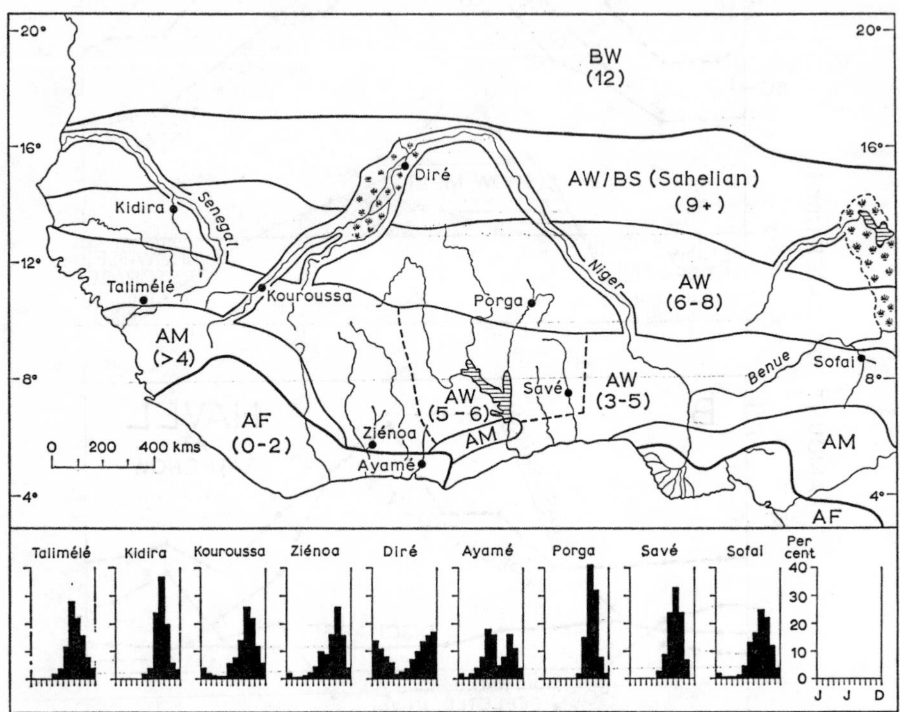

Fig. 10.1.2 Hydrological regions and characteristic river regimes of West Africa (Adapted from Ledger, 1964 and *International Atlas of West Africa*, Plate 10).

Thick solid line denotes approximate boundary of major hydrological region; thin solid lines show subdivisions of major region; thin pecked line shows minor variant of a subdivision, in this instance AW (5–6) or Dahomean sub-type. Numbers denote approximate length in months of dry season. Symbols are explained in the text. BS (arid steppe) and BW (desert) are areas where permanent streams cannot originate.

floodplain rivers contract rapidly to their low-water channel, leaving large areas of bed exposed. However, many rivers, especially those flowing on fine alluvium, tend to contract relatively little, because their channels are naturally (and in places artificially) puddled with clay particles. But for this puddling, most riverine plains near sea-level would be constantly waterlogged or flooded, as they lie well below mean river-level.

The regime of a river will also be affected by the geometry of its drainage basin. Thus high convex landforms in the tropics and sub-tropics favour the formation of local orographic uplift cells during spells of intense daylight insolation which often cause heavy convectional rainfall on the upper mountain slopes. This is a frequent cause of violent spates in mountain torrents on tropical highlands and islands.

Channel characteristics, such as size and shape of the cross-section and slope and roughness of bed, are partly the result of the regime, and have only a minor influence upon it. On the other hand, the morphometry of a basin, which involves the size, shape, stream-pattern, and orientation of a drainage basin, has a more decisive influence upon both the regime and its speed of reaction to climatic factors. Very small basins tend to show rapid reactions and violent characteristics, to be hypersensitive to brief downpours, during which overland flow is more important than channel flow. As basin size increases, the channel storage effect becomes increasingly dominant. In sizeable drainage basins the basin shape and the stream pattern may either modulate or accentuate the regime. This is largely a question of the convergence or non-convergence of tributaries and of the coincidence or non-coincidence of times of arrival of flood and of low-water. The total length of channel is also significant, as it affects the local arrival of seasonal variations from upstream. This is well shown on the Niger, where the upstream high water (September–October), after travelling over 2,000 km largely across a flat swampy plain, forms a flood on the middle course from January to March (fig. 10.1.2).

2. Types of river regimes, or hydrological regions

The above discussion reveals the importance of the size of units in the analysis and classification of river regimes. Whereas the river regimes of small and moderately sized basins may closely reflect regional runoff controls, especially climate, the main watercourses of many large and complex basins often acquire regimes unrepresentative of the territory they are crossing. The lower Colorado and lower Nile are obvious, and the lower Rhine and lower Rhône less obvious examples. There are, however, such large areas of the world within which local and regional river regimes reflect the regional climatic rhythm that some form of areal differentiation into river-regime types or *hydrological regions* seems desirable. Notable attempts at identifying hydrological regions have been made recently, particularly for Italy, France, and West Africa (fig. 10.1.2). This areal differentiation will progress in accuracy and coverage as hydrological observations increase, and will, without the need for extrapolation, eventually be based entirely on local measurements, showing the influence of both regional climatic and non-climatic factors on runoff. It will also be possible to distinguish all rivers with regimes markedly different from that of the hydrological region they are crossing.

At present these complexities are shown by inserting on maps hydrographs for selected points (fig. 10.1.3). Such a method is excellent where stations abound, but these data for a station on a river represent more than the discharge at that location; they relate also to the hydrological area providing the runoff or the total environment that a wise engineer would not ignore. Thus on large-scale maps station regime hydrographs *and* hydrological regions need to be inserted. As the scale of the map decreases, however, it becomes more and more difficult to show sufficient hydrographs and increasingly convenient to equate characteristic river regimes with generalized hydrological regions. If these regional

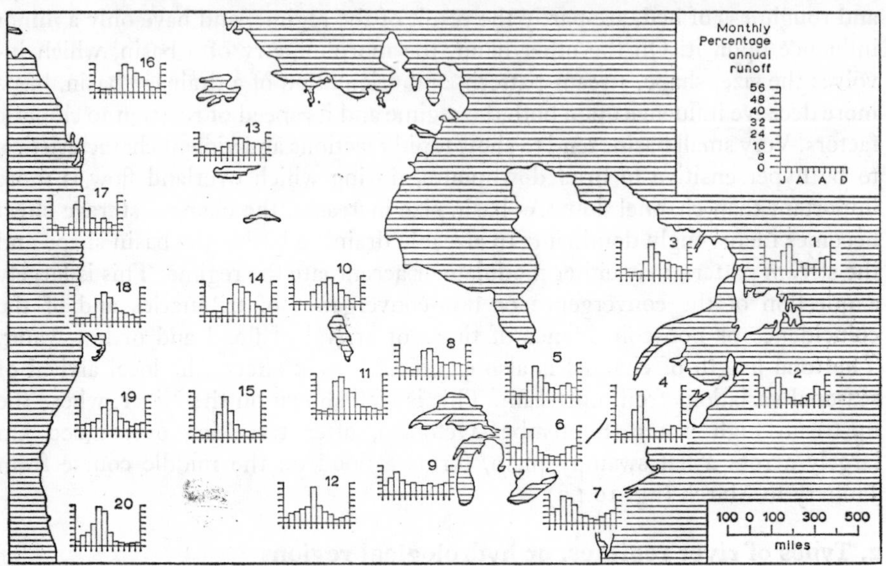

Fig. 10.1.3 Regime hydrographs of typical rivers in Canada and the northern United States (After Bruce and Clark, 1966 and Langbein and Wells, 1955).

General hydrological regions are given in Fig. 10.1.4. The rivers are: 1. Gander R., Newfoundland; 2. St. Mary's R., Nova Scotia; 3. Hamilton R., Labrador; 4. Upsalquitch R., New Brunswick; 5. Harricanaw R., Quebec; 6. Saugeen R., Ontario; 7. Susquehanna at Harrison, Pa; 8. English R., Ontario; 9. Pecatonica R., Illinois; 10. Saskatchewan R., Manitoba; 11. Assiniboine R., Manitoba; 12. Republican R., Nebraska; 13. Yellowknife R., N.W.T.; 14. N. Saskatchewan R., Edmonton, Alberta; 15. Yellowstone R., Montana; 16. Yukon R., Dawson; 17. Skeena R., Usk, B.C.; 18. Fraser R., Hope, B.C.; 19. Columbia R., The Dalles, Oregon; 20. Kings R., Piedra, California.

regimes are designated by a shorthand nomenclature (initials) it is possible to indicate the following, even on small-scale maps:

1. the boundaries of the hydrological regions;
2. the nature and components of the river regimes, however complex; and
3. any special local influences, quite apart from regional ones.

The climatic terminology used by Köppen seems adaptable for this purpose and is desirably genetic. In the world classification used below it is assumed that Köppen's terms retain their climatic meaning:

A = tropical rainy climates; all months with mean of over 18° C.

B = dry climates with an excess of potential evapotranspiration over precipitation.

C = warm, temperate rainy climates.
D = cold, snowy climates; the mean temperature of the coldest month being not more than −3° C.

The rainfall symbols of Köppen are applied directly to runoff and are promoted to be the second capital letters in the shorthand, thus F denotes appreciable runoff all the year. W marked winter low-water, and S summer low-water. It will be noticed that the major hydrological regions of the world (fig. 10.1.4) show a remarkable general coincidence with Köppen's climatic divisions. Where discrepancies occur they are in part due to the need for revising Köppen's scheme in the light of modern climatic statistics (as in the Amazon basin) and as the result of modern knowledge of the distribution of vegetation types. In fig. 10.1.4 the boundaries of many of the divisions will remain tentative until more hydrological details are available. The possibility of further subdivision of larger hydrological regions is illustrated in fig. 10.1.2. An attempt is also made in fig. 10.1.4 to distinguish between tropical areas where the river regimes have no marked low water (AF) and those where the tropical rivers experience an appreciable low-water season (AM), but which is not sufficiently severe nor sufficiently prolonged to allow them to be classified as AW.

The third category of letters designates (in small type) temperature regimes, which obviously also have relevance to hydrological regimes:

a = Mean of warmest month over 22° C.
b = Mean of warmest month under 22° C; more than four months averaging over 10° C.
c = One to four months averaging over 10° C, and mean of warmest month under 22° C.
d = Mean of coldest month under −38° C.

A. Megathermal regimes (A)

1. AF: EQUATORIAL DOUBLE MAXIMA. In some equatorial areas the heaviest rains occur in spring and autumn following the equinoxes, and there is no dry season. Regions exhibiting these double maxima are the main valley of the Congo lying athwart the equator, probably parts of Indonesia and of the upper Amazon, and a coastal strip in eastern Brazil. In these cloudy areas the surface insolation is rarely excessive and allows the abundant rainfall to maintain a high discharge all the year; however, the coastal parts of West Africa that experience this regime seem to have a brief but marked low-water stage.

The widespread invasions of summer maritime (*mT*) air masses give much of the tropics and sub-tropics a strong summer maximum of runoff. After the rains the rivers dwindle to a marked minimum in late winter and early spring when insolation is great. The duration of high water varies with the length and intensity of the rainy season, and as a rule decreases rapidly inland, except where rising relief intervenes. Probably several hydrological subdivisions are neces-

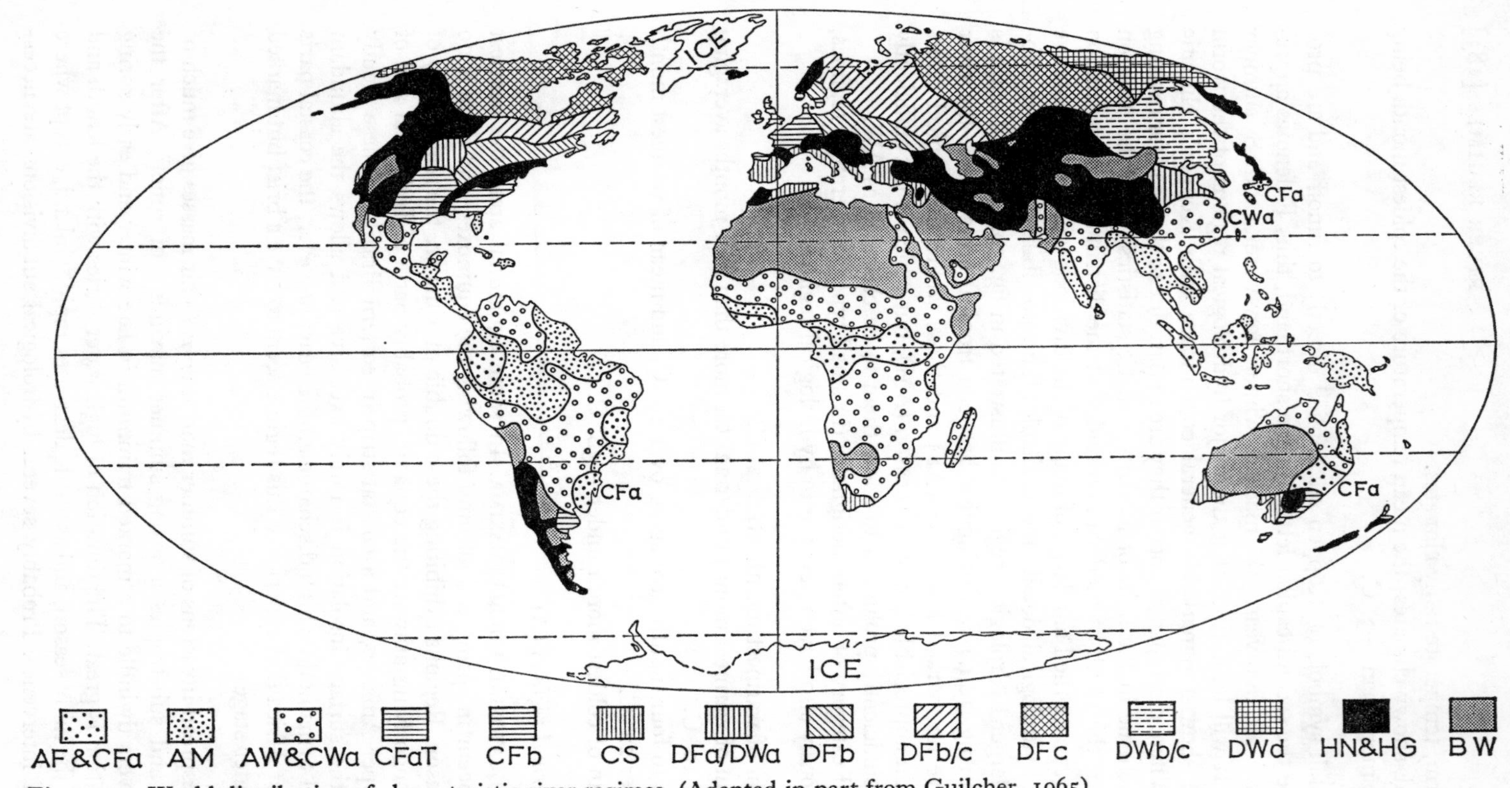

Fig. 10.1.4 World distribution of characteristic river regimes, (Adapted in part from Guilcher, 1965).

Symbols are explained in the text and illustrated in Figs. 10.1.5 and 10.1.6. BW denotes desert and other dry areas (BS) where streams cannot originate.

Mountain zones in some areas of Asia and the western United States should be shown as watercourses crossing desert (BW) and dry steppe (BS) regions, as is done for southern South America, but such details could not be attempted at this scale.

sary, based on the length of the low water (figs. 10.1.2 and 10.1.5), but information is usually inadequate for this at the present. Suggested divisions are:

2. AM: TROPICAL STRONG SINGLE MAXIMUM WITH A SHORT LOW WATER PERIOD OF UNDER THREE OR FOUR MONTHS. The rainier parts of West Africa and vast areas of the Amazon basin and of monsoonal South-East Asia belong to this heavy rainfall type.
3. AW: TROPICAL SINGLE MAXIMUM WITH A LONG LOW WATER. Where the

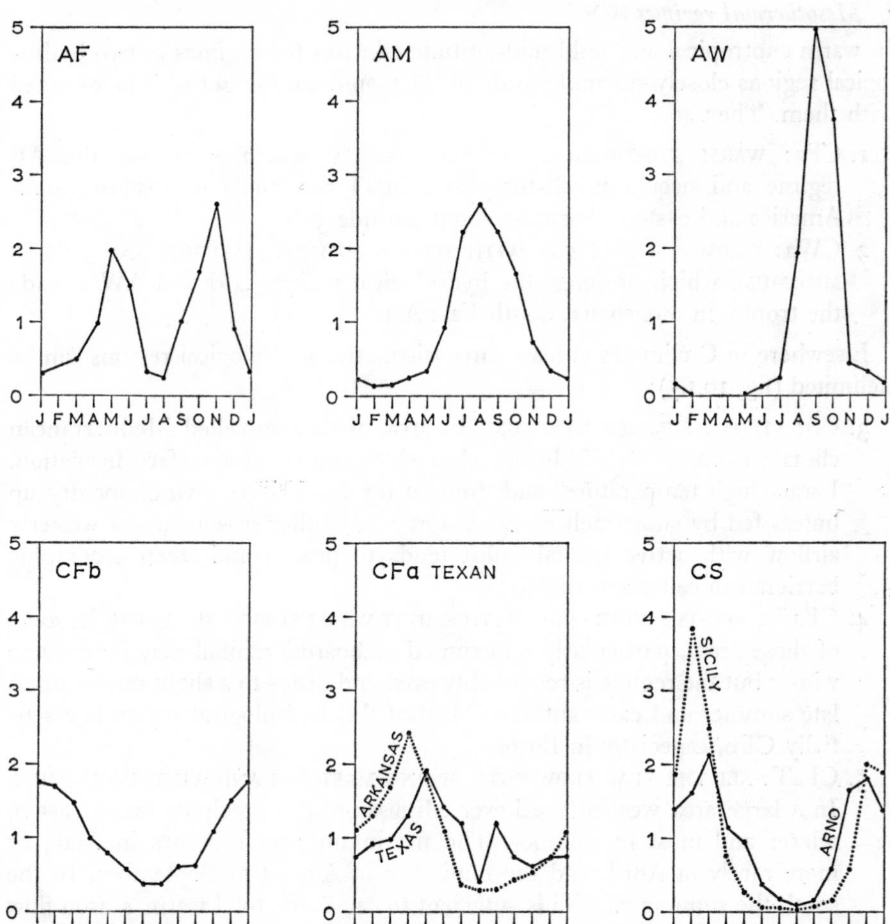

Fig. 10.1.5 Characteristic river regimes controlled mainly by rainfall and warm season evaporation.

AF. Lobaye R., a northern tributary of the Congo; AM, Lower Irrawaddy; AW, Pendjari R., a tributary of the Volta (See Fig. 10.1.2); CFb, Thames R., England; CFa Texan Buffalo R., Arkansas and Guadalupe R., Texas; CS. Arno R. and Imera Meridionale, Sicily.

All graphs show monthly coefficient of mean monthly flow for whole year.

rainy season shortens and annual falls lessen, the low-water period increases from four or five to six or seven months, and most small rivers become almost or quite dry. With increasing aridity the dry season extends to eight or nine months or more, and the hydrological characteristics degenerate into semi-arid steppe where permanent streams cannot originate and favoured watercourses have an episodic flood only once or twice a year, and ultimately into desert (BW), characterized by very infrequent flash floods.

B. Mesothermal regimes (C)

In warm subtropical and mild mid-latitude climates the regimes in two hydrological regions closely resemble those in the tropics and might well be grouped with them. They are:

1. CFa: WARM SUB-TROPICAL DOUBLE MAXIMA, which resembles the AF regime and occurs in all-the-year rainfall coastlands of eastern South America and eastern Australia about latitude 30°.
2. CWa: WARM SUB-TROPICAL WITH STRONG SUMMER MAXIMUM AND WINTER MINIMUM, which prolongs the hydrological regions AM and AW outside the tropics in monsoonal South-East Asia.

Elsewhere in C climates at least three distinctive hydrological regions can be delimited (fig. 10.I.5):

3. CS: STRONG SUMMER MINIMUM. In areas with a so-called Mediterranean climate summer usually brings clear skies and intense surface insolation. Under high temperatures and drought the local rivers dwindle or dry up unless fed by snow-melt or karst storage. In other seasons moist westerly airflow with active frontal uplift tends to prevail and steep orographic barriers can cause severe floods.
4. CFa/b: ALL-THE-YEAR FLOW WITH SLIGHT WARM-SEASON MINIMUM. In some of these areas, particularly on exposed seaboards, rainfall may be most in winter but the regime is remarkably even and drops to a slight minimum in late summer and early autumn. Most of this hydrological region is essentially CFb, especially in Europe.
5. CFaT: ALL THE YEAR FLOW WITH SPRING MAXIMUM AND WINTER MINIMUM. In a large area west of the lower Mississippi the precipitation is least in winter and most in summer. The maximum runoff occurs in May, or more rarely in April, and the minimum in August or September. In the south the summer rainfall is sufficient to cause the total warm season flow to exceed slightly that of winter, but farther north these conditions are reversed. Perhaps in the classification suggested it would be better to replace T, denoting Texas the typical regional location, by the number of the calendar month of maximum flow—i.e. CFa^5 or CFa^4.

In the three hydrological regions just described snow falls occasionally in the coldest months, but, except on hills, it seldom lasts more than a few consecutive days, and very seldom causes a flood, except in early spring.

C. Microthermal regimes (D)

Where one or more months experience mean temperatures below −3° C, snow-cover normally lasts for a month or more. As the frigidity of the winter half-year increases so does the proportion of the cold-season precipitation that falls as snow and is lost to the winter runoff and added to the spring flood. The actual depth of snow tends to decrease, and the depth and duration of freezing of soils and rivers to increase, with the severity of winter temperatures. Everywhere most precipitation falls in summer, and the annual totals are mediocre or small, being usually under 1 m in eastern Canada, 600 mm in Europe, and 400 mm in Siberia. Six main hydrological regions can be distinguished, three of which occur mainly in eastern Asia (fig. 10.1.6):

1. DFa/DWa: SUMMER PLUVIAL MAXIMUM; WINTER NIVAL MINIMUM. In north China and in the vicinity of the state of Kansas in North America the river regimes show a marked summer maximum (coincidental with the pronounced summer rains) and a winter minimum due to relatively cold, dry, snowy weather. Locally, snow-melt causes a brief secondary maximum in spring.
2. DWb/c: STRONG SUMMER PLUVIAL MAXIMUM; LONG WINTER NIVAL MINIMUM. North of DWa regions, in the eastern Asia coastlands draining mainly to the Sea of Okhotsk, the frost-bound period lasts for six or seven months. On rivers such as the Amur the snow-melt in May or June causes a moderate flood which, often after a slight recession, rises in August or September to a main maximum due to summer rains (fig. 10.1.6).
3. DWd: STRONG SUMMER PLUVIO-NIVAL MAXIMUM: PROLONGED COLD SEASON MINIMUM. In north-eastern Siberia the total annual precipitation averages well under 250 mm, and in parts under 100 mm. The maximum is strongly concentrated in summer, as the severe winters allow very little snowfall. Permafrost remains near the surface all the year, and in winter most rivers freeze solid except in deep pools, so that cold-season runoff is practically nil. With the thaw in June the flood is moderate (although truly excessive compared with the negligible winter runoff) and merges directly into the runoff from the July–August rains to form then a single pluvio-nival maximum.

In the colder parts of Europe and of Asia and North America not described above the hydrological regimes are dominated by winter snow and by summer evapotranspiration, although maximum rainfall occurs in summer. Vast areas were affected by thick ice sheets during the Quaternary glaciations and have an immature drainage, with numerous lakes and swamps that greatly moderate the warm-season flow. Three major characteristic regimes are distinguishable:

4. DFa/b: MODERATE PLUVIO-NIVAL OR NIVO-PLUVIAL SPRING MAXIMUM; SLIGHT SUMMER MINIMUM. In much of New England and the southern parts of the St Lawrence basin, and in a broad belt from southern Sweden to the

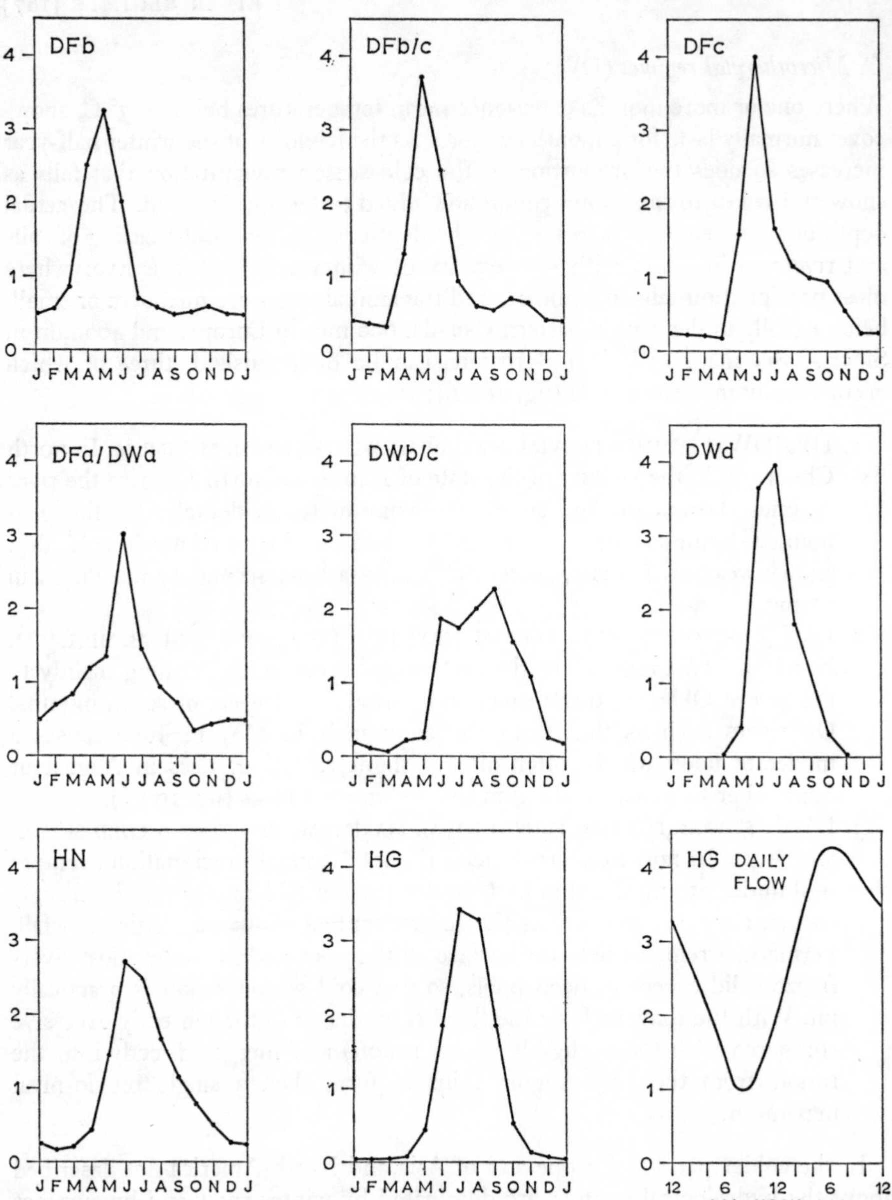

Fig. 10.1.6 Characteristic river regimes controlled mainly by cold season snowfall and warm season rainfall.

DFb Dnepr R. at Kremenchug; DFb/c Volga R. at Kuybyshev; DFc Yenisey at Igarka; DFa/DWa Republican R. near Bloomington, Nebraska; DWb/c Amur R. at Komsomol'sk; DWd Indigarka R.; HN Reuss R. at Andermatt; HG Massa R. at Massaboden. HG Daily Flow, South Cascade Glacier stream during a fine warm spell (After Meier and Tangborn, 1961).

All graphs, except the last, show monthly coefficient of mean monthly flow for year.

Black Sea in Europe east of the Elbe basin, the river regimes are either pluvio-nival or nivo-pluvial. In the former a moderate spring spate (March or April) is followed by a slight low water in August or September. In the latter the spring maximum is slightly stronger and comes in April or May, while the main winter flow may be only slightly greater than the summer minimum. Most of this region is definitely DFb, particularly in Europe.

5. DFb/c: STRONG NIVAL SPRING MAXIMUM; SECONDARY AUTUMN PLUVIAL MAXIMUM. Over large areas in eastern Canada, European Russia, and northern Scandinavia a variant of the nivo-pluvial regime dominates. A strong snow-melt maximum (usually in May) directly follows a cold-season minimum (January–March) and is itself followed by a small secondary maximum in late autumn (October–November) when a moderate rainfall is not offset by high evapotranspiration (figs. 10.1.3 and 10.1.6).
6. DFc: VIOLENT NIVAL SPRING MAXIMUM; STRONG WINTER MINIMUM. In northern Canada and the northern expanses of western and central Siberia (draining to the Arctic Ocean by the lower courses of great rivers such as the Ob, Yenisey, and Lena) a classic lowland nival regime prevails. A severe low water lasts from December to late April or early May, during which a small discharge persists beneath the carapace of ice on rivers and lakes. In May and June a sudden thaw occurs simultaneously over wide areas and causes a violent June maximum. The mean monthly flow of the lower Yenisey at Igarka in June (78,000 m^3/sec) is exceeded only on the Amazon. The decline of the flood is less rapid, as it is moderated slightly by the effect of summer and early autumn rains. A feature of the northward-flowing Siberian rivers is that the upper and middle courses usually thaw out between late April and mid-May, whereas their mouths remain frozen until early June. The resultant floods spread out over wide areas and floating ice menaces banks and structures.

D. Mountain regimes (Hohenklima: above 1,500 m)

Widespread forms of microthermal hydrological regimes occur on high mountains outside the polar ice-caps, and it is proposed to designate these HN (*nivale* or Highland Snow) and HG (*gletscher* or Highland Ice). Because mean shade temperatures decrease upwards on an average about 6° C per 1,000 m, there is an elevation on most high mountains when precipitation begins to accumulate as snow. The margin of the snow-cover is lowest in winter, except in the tropics, where it lowers in the wet season. In the warm season the snow melts back and, where the height is sufficient, retreats to a permanent (*firn*) snowline. During this ascent the melt-water spate of the rivers lasts as long as the snow cover. Where ice forms, it persists as glaciers, sometimes far below the permanent snow-line, and ice-melt in summer may long sustain a river's flow. On lofty tropical and sub-tropical mountains the thin atmosphere allows on clear high-sun days intense solar radiation which causes a rapid melt-water runoff. Such streams often grow daily to a late-afternoon flood and dwindle fast after nightfall as temperatures fall rapidly below freezing-point (fig. 10.1.6). Travellers should cross them at first light, thereby avoiding the afternoon spate and the chance in hot weather of flash floods due to convectional downpours near the summits.

The mountain slopes between the lower and upper seasonal positions of the snowline may experience rain in the warmer months. Pardé has classified regimes with an appreciable snow-cover influence on the basis of a *coefficient of nivosity* which expresses the percentage of the warm-season flow contributed by melt-water. For Alpine mountains in Switzerland and Savoy the coefficient was the basis for the following subdivisions:

6–14%: pluvio-nival
15–25%: nivo-pluvial
26–38%: transition to nival
29–50%: pure nival to nivo-glacial
51% and over: glacial (rising to 67% on the Massa basin which is nearly 70% ice-covered).

Probably the enormous snowfalls on mountains in western North America would yield higher coefficients of nivosity than occur in Alpine Europe. But irrespective of the critical statistics used, it seems that these lower mountain slopes must be considered as a gradation of DFa/b (pluvio-nival), DFb (nivo-pluvial), and DFb/c (transition to nival) lowland regimes above a CFa/b base. However, the HIGHER slopes can hardly be considered DFc and DFd, as at some height the thinness of the mountain atmosphere introduces the unique excessive diurnal range during high-sun periods. As a result of rapid night cooling, above the firn-line the main daily temperatures normally remain below zero all the year. Even on very snowy mountains where large snowfalls may persist when mean summer temperatures are well above zero, frosts can still be expected at night. Because of this great diurnal range and proneness to night frost, the pure nival (or nivo-glacial) and the glacial hydrological regions on mountains should be designated by HN and HG respectively.

Since the effect of the climate on HN and HG river regimes depends on the presence of snow or ice, it is difficult to give altitude limits to these hydrological regions, except for broad latitudinal zones. There are, however, great differences between the effect of mountain-snow in the tropics and in high latitudes. In the tropics the temperature difference between the seasons is relatively small and the high basal temperatures ensure a high snowline all the year. Thus seasonal snow-melt areas are small and the local firn-line (HG) lies at about 4,600–6,000 m. Away from the topics the seasonal temperature differences rapidly increase, and both the permanent and temporary snowlines rapidly lower, the latter being at or near sea-level in D climates. In the same way the area affected by warm season snow-melt on mountains also greatly increases.

In most sub-tropical and cool-latitude mountains the deeply dissected relief and wide variations on local snowfall cause the HN and HG hydrological regions to be intricately interconnected. In the western Alps the mean altitude of the nivo-glacial basins varies from about 1,400 to over 2,000 m. The HG regions are usually situated above these altitudes, and commonly have 15–20% or more of their basins ice-covered. Here the river regimes normally show a single summer maximum culminating in July (or rarely in June), whereas throughout nivo-glacial regions the rivers commonly reach their maximum in June (fig. 10.1.6). These slight variations in peak flow, as with those of any lowland regime, could be indicated by adding the calendar number of the month of high water to the symbols. Thus HG^6 equals June and HN^5 May. In the southern hemisphere HG^1 (January) and HG^2 (February: ultra-glacial) occur in New Zealand and southern Chile. Worth indicating also is the liability of HG regions to glacial melt-water debacles, for which the symbol j (from the Icelandic *jokulhlaup*) seems suitable.

Indeed, for all regimes the genetic shorthand nomenclature could, where necessary, be made more explicit by adding the calendar month of high-water, in addition to symbols for special local influences such as:

- j. jokulhlaup
- l. moderated by lake storage
- p. moderated by porous catchment
- v. retarded by vegetation in channel, and
- i. lowered by irrigation abstraction.

Acknowledgement. The author is especially indebted to Professor Maurice Pardé for great help and encouragement over a long period of time.

REFERENCES

Excellent summaries with good bibliographies are:

GUILCHER, A. [1965], *Precis d'Hydrologie, marine et continentale* (Masson, Paris), pp. 267–379 (Also with good section on lakes.)

KELLER, R. [1962], *Gewässer und Wasserhaushalt des Festlandes* (Teubner, Leipzig), 520 p.

PARDÉ, M. [1949], *Potamologie* (University of Grenoble) (Roneographed), 2 vols., 336 p.

PARDÉ, M. [1955], *Fleuves et Rivières* (Colin, Paris), 3rd edn., 224 p.

PARDÉ, M. [1961], *Sur la puissance des crues en diverses parties du monde* (Geographica, Saragossa), 293 p.

WUNDT, W. [1953], *Gewässerkunde* (Berlin and Heidelberg), 320 p.

Excellent shorter summaries, also with bibliographies are:

BRUCE, J. P. and CLARK, R. H. [1966], *Introduction to Hydrometeorology* (Pergamon, London) (especially pp. 33–56).

CHOW, V. T., editor [1964], *Handbook of Applied Hydrology* (McGraw-Hill, New York), 1,418 p. (Section 14 on 'Runoff', by Ven Te Chow, and Section 16 on 'Ice and Glaciers', by Mark F. Meier.)

ROCHEFORT, M. [1963], *Les Fleuves* (P.U.F., Paris) ('Que sais je?'), 128 p.

AF, *AM*, *AW and CFa Regimes.*

Among regional monographs and articles are:

HURST, H. E. [1952], *The Nile* (Constable, London), 326 p.

LEDGER, D. C. [1964], Some hydrological characteristics of West African rivers; *Inst. Brit. Geogr.*, 35, pp. 73–90.

LOCKERMANN, F. W. [1957], *Zur Flusshydrologie der Tropen und Monsunasiens* (Bonn) (Roneographed), 619 p.

ROCHEFORT, M. [1958], *Rapports entre la pluviosite et l'écoulement dans le Brésil subtropical et le Brésil tropical Atlantique* (Paris), 279 p.

RODIER, J. [1963], *Bibliography of African Hydrology* (UNESCO, Paris), 166 p.

RODIER, J. [1964], *Régimes hydrologiques de l'Afrique Noire à l'Ouest du Congo* (Paris).

C and D Regimes

IONIDES, M. G. [1937], *The Régime of the Rivers Euphrates and Tigris* (Spon, London), 278 p.

MASSACHS ALAVEDRA, V. [1948], *El regimen de los rios peninsulares* (Barcelona), 511 p.

PARDÉ, M. [1964], Les régimes fluviaux de la péninsula Ibérique; *Revue de Géographie de Lyon*, 129–82.

For *Italy* there is an excellent map in:

A. R. TONIOLO, *Atlante Fisico Economico d'Italia*, 1940, Map 9, Idrografia Terrestre.

Europe: annual review of hydrological literature in:

Revue de Geographie de l'Est, (Nancy), 1961, onwards by RENÉ FRÉCAUT.

U.S.A.:

LANGBEIN, W. B. and WELLS, J. V. B. [1955], The water in the rivers and creeks; In *Water*, U.S. Department of Agriculture Yearbook, pp. 52–62.

LANGBEIN, W. B. *et al.* [1949], Annual runoff in the United States; *U.S. Geological Survey Circular* 52.

Glacial: HG Regimes

MEIER, M. F. and TANGBORN, W. [1961], Distinctive characteristics of glacier runoff; *U.S. Geological Survey Professional Paper* 424.

For further references to debacles, see:

MEIER, M., In Chow V. T. [1964], Section 16, pp. 30–2. and

BECKINSALE, R. P. [1966], *Land, Air and Ocean;* 4th edn. (Duckworth, London), pp. 327–8 and 339.

Vegetation

CHOW, V. T. [1959], *Open-channel Hydraulics* (McGraw-Hill, New York), pp. 102–5.

WARD, R. C. [1965], Evapotranspiration from the Thames floodplain; in Whittow, J. B. and Wood, P., Editors, *Essays in Geography* (University of Reading), pp. 145–67.

11.1. Long-term Precipitation Trends

R. G. BARRY

Institute of Arctic and Alpine Research, University of Colorado

In Chapter 3.1(i) some indication was given of the variability of precipitation amounts, but the question of long-term trends was not explicitly discussed. This is clearly of relevance to our consideration of geomorphic processes and man's water requirements. Direct study of precipitation changes is unfortunately very restricted by the limited availability of precipitation records exceeding 100 or even 50 years duration, and it is vitally important, therefore, that such information as is available be analysed and interpreted correctly. For this reason we begin with a summary of the terminology of climatic change and of the simpler methods of analysing time series. Fuller details may be found in Mitchell *et al.* [1966], which provides the basis for the following account.

1. The terminology of climatic change

Three basic types of change can be distinguished. They are: a *discontinuity* – an abrupt and permanent change in the average value; a *trend* – a smooth increase or decrease, not necessarily linear, of the average; a *fluctuation* – a regular or irregular change characterized by at least two maxima (or minima) and one minimum (or maximum). Where a climatic fluctuation progresses smoothly and gradually between the maxima and minima it is termed an *oscillation*, and if the maxima and minima recur after approximately equal time intervals it is referred to as a *periodicity*.

It will be noted that these definitions do not incorporate any reference to time scale. Changes occurring during the instrumental period, on a scale of 10–10^2 years, are termed '*secular*', those of the order of 10^3 years '*historical*'.

2. Time series

The year-to-year variability of precipitation totals may conceal long-term changes of one kind or another in a data series, and statistical techniques are necessary to suppress the short-term irregularities. The simplest method is the calculation of a *running mean* (or *moving average*), where mean values are determined for successive, overlapping periods of five, ten, or thirty years. For example, in the five-year case

$$\frac{P_1 + P_2 + P_3 + P_4 + P_5}{5} = \bar{P}_3$$

$$\frac{P_2 + P_3 + P_4 + P_5 + P_6}{5} = \bar{P}_4$$

where P_1 = precipitation in year 1 of the series;

$\bar{P}_3$ = running mean value for year 3.

More generally,

$$\bar{P}_x = \sum_{i=-n}^{n} P_{x+i}/2n + 1 \qquad (1)$$

where $\bar{P}_x$ = running mean value for x^{th} term of the series;

$2n + 1$ = number of terms in the running mean;

$\sum_{i=-n}^{n}$ = summation of the terms from $i = -n$ to $i = n$.

Figure 11.1.1 illustrates the smoothing effect of various running means for annual precipitation totals at Omaha, Nebraska. In the case of the five-year series there is a tendency for displacement and apparent inversion of the shorter fluctuations, but in general the effect is to clarify the long-term changes. There

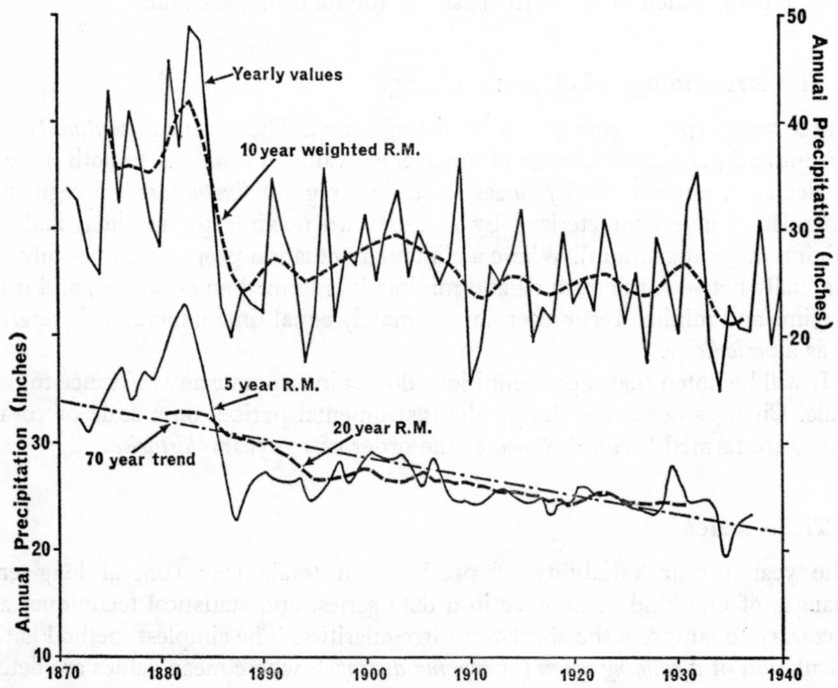

Fig. 11.1.1 Annual precipitation at Omaha, Nebraska, 1871–1940, and the smoothing effects of running means (Based on Foster, E. E., *Rainfall and Runoff*, 1949).

is a risk that some of the prominent longer fluctuations in the smoothed series arise from random variations, and indeed an original series of wholly random data can show quasi-cyclical rhythms after smoothing (Lewis, 1960). Simple statistical tests are available to assist the interpretation of smoothed series (Craddock, 1957).

Modification of this smoothing technique can give better results and avoid the inversion of the shorter fluctuations, noted above. For example, it is often useful to use a weighted running mean. This is termed a *low-pass filter*, since it retains changes of long wavelength and filters out the short ones. The procedure for determining appropriate weights cannot be dealt with here, but the following table illustrates a set of weights for studying variations with a wavelength longer

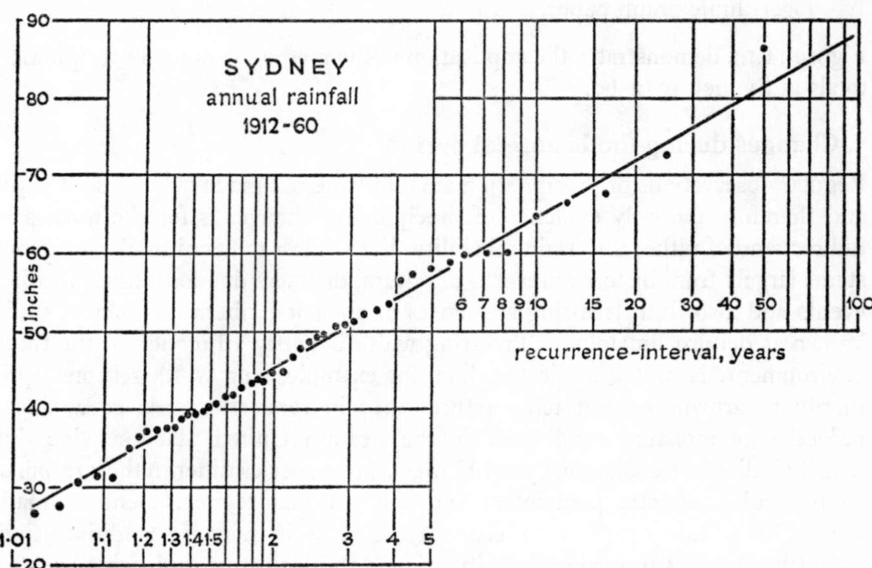

Fig. 11.1.2 Annual precipitation at Sydney, Australia, 1912–60, illustrating the calculated return periods (From Dury, 1964).

than 10 terms in the data series. The weights are based on the Gaussian (Normal) frequency distribution.

$i =$	−4	−3	−2	−1	0	1	2	3	4	Sum
Weight =	0·01	0·05	0·12	0·20	0·24	0·20	0·12	0·05	0·01	1·00

Figure 11.1.1 shows the application of this filter to the Omaha precipitation data. It is clear that this procedure gives more satisfactory smoothing than the simpler unweighted running means. The amplitude and time-phase of the fluctuations are accurately represented.

So far we have been concerned only with series of average values. Often, a more important question is the frequency of extreme values. A useful, basic technique is the analysis of *return period* (or recurrence interval); that is, the average time interval within which a rainfall, or flood, of specified amount or intensity can be expected to occur once. For example, a 10-year annual rainfall is the probable annual maximum once in a decade and the 100-year annual rainfall once in a century. This does not imply that the spacing is regular. A series of 70 years' data may well include the 100-year maximum and even the 500- or 1,000-year maximum. The method is as follows:

1. Rank the n observations in order of magnitude with the highest as 1 and the lowest as n.
2. The return period $= (n + 1)/r$, where r is the rank of a particular observation.
3. Plot the actual values against their computed return periods on semi-logarithmic graph paper.

Figure 11.1.2 demonstrates the application of the method to annual precipitation totals at Sydney 1912–60.

3. Changes during the historical period

Regular observations of precipitation amount were not made before about 1850, and therefore the only evidence of precipitation changes is for the most part indirect and of rather uncertain reliability. For the last millennium this evidence stems largely from historical records of natural disasters due to unusual weather events and also from tree-ring sequences of ancient timbers – a field of study known as dendroclimatology. Tree-ring width is partly a function of the tree's environment. Near the arctic tree-line, for example, ring width responds primarily to growing-season temperatures, but in semi-arid areas growth is a reflection of moisture conditions. In the western United States a wide ring usually indicates a cool, moist year. However, tree species differ in their response to particular climatic parameters and the seasonal regime. Dendroclimatic studies have made it possible to reconstruct changes in moisture conditions since the fifth century A.D. for locations in Arizona and Colorado, and Fritts has prepared maps of regional patterns of moist, cool or dry, warm decades since 1500 in western North America. More tentative indications of moister and drier periods during the last 10,000 years may be gleaned from analysis of peat-bog stratigraphy, especially pollen profiles preserved in the bogs. This evidence, like that of tree rings, is indicative of the moisture budget P-E, rather than precipitation alone. Moreover, the bog stratigraphy may reflect the local hydrological regime rather than climatic changes. For this reason we shall limit the discussion here to the surer documentary material.

Studies by Lamb [1965] of the historical records of good or bad harvests, of floods or droughts, and so on provide an index of 'summer-wetness' (July–August) in Europe since A.D. 1100 (fig. 11.1.3(a)). Each month scores 0 for drought evidence, $\frac{1}{2}$ if unremarkable, or 1 for indications of wetness. Decadal extremes of the index are 4–17, with a value of about 10 for unremarkable ones. A more tentative extrapolation has been made for the three centuries back to A.D. 800. Figure 11.1.3(b) shows the 50-year average rainfall as a percentage of

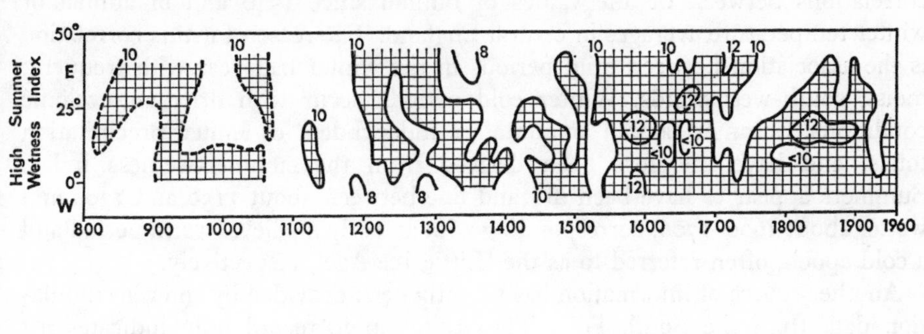

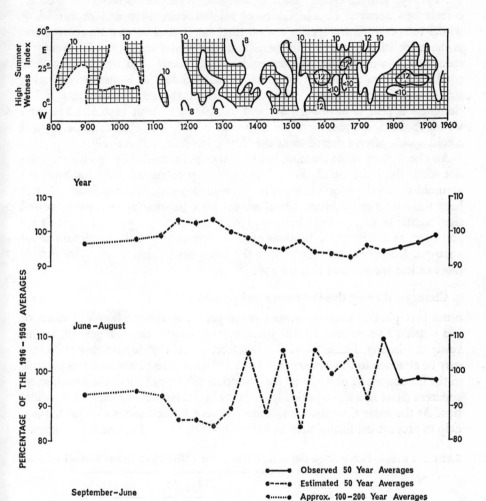

Fig. 11.1.3 Precipitation trends in Europe since A.D. 800 (After Lamb, 1965). See text for details.
(*Top*). 'Summer-wetness' (July–August) in Europe near 50° N since A.D. 800.
(*Below*). Fifty-year averages of rainfall in England and Wales expressed as percentages of the mean for 1916–50.

the 1916–50 annual average for England and Wales. This series is based on correlations between decade values of rainfall since 1740 and of annual or winter temperature averages in central England. The reason for this correlation is the association between mild periods in the winter half-year with frequent, moist, south-west winds. Winter cold periods occur with drier anticyclonic conditions. Summer rainfall amounts are independent of annual precipitation totals, and have therefore been derived from the summer-wetness index. Summers appear to have been dry and fine between about 1150 and 1300 and wetter about 1600–1700, corresponding with an early medieval warm period and a cold epoch, often referred to as the 'Little Ice Age', respectively.

Another source of information has recently been provided by snow accumulation data from the South Pole. The stratigraphic record only indicates *net* accumulation – the effects of wind deflation or deposition are unknown – but at least the evidence is direct. Mean annual accumulation apparently increased significantly from 5·4 g cm^{-2} between 1760 and 1825 to 7·5 g cm^{-2} between 1892 and 1957, and there are indications of some decrease since the maximum about 1920–30. Deeper layers suggest that the mean accumulation for 1550–1750 is more or less the same as that for 1760–1957.

4. Changes during the instrumental period

Since precipitation measurements cover a period of about a hundred years, or less outside Europe and North America, only short-term changes can be investigated in any detail. There is, therefore, a risk that too much significance may be attached to these changes. They need to be interpreted as far as possible in the framework of post-glacial and historical climatic change. For example, an apparent trend in a short-period record may in fact be part of a long-term oscillation. At the same time the complexities of recent fluctuations (Veryard, 1963) help to prevent the formulation of over-simplified views about earlier changes.

TABLE 11.1.1 Percentage departure from the 1881–1940 mean annual rainfall

Station	1874–98	1907–31
Barbados, West Indies 13° N	14	−9
Honolulu, Hawaii, 21° N	13	−12
Townsville, Queensland, 19° S	17	−4
Colombo, Ceylon, 7° N	4	−8
Freetown, Sierra Leone, 8° N	11	−12

In the tropics there is evidence of a marked decrease of annual rainfall over extensive areas at the end of the nineteenth century. This is illustrated by Table 11.1.1 from Kraus [1955]. The greater aridity in the tropics in the early twentieth century was matched by rainfall decrease in south-east Australia and eastern North America, pointing to a general weakening of the moisture cycle in lower latitudes. Work in the Mediterranean area by Butzer [1957] strengthens this impression. Figure 11.1.4 shows that the Saharan–Arabian desert and the

Mediterranean region became much drier in the present century, whereas the North Atlantic, north-west Europe, and southern Russia experienced an increase of annual precipitation until about 1940. There is evidence that the trend towards drier conditions in the tropics was reversed during the 1930s, and in eastern North America and eastern Australia there has been a post-1940 increase in rainfall, apparently caused by more frequent tropical cyclones. It is significant that the changes have affected areas close to climatological boundaries, whereas the climate in 'core areas' such as Antarctica and the equatorial rain-forest seems to have remained more or less constant over long periods.

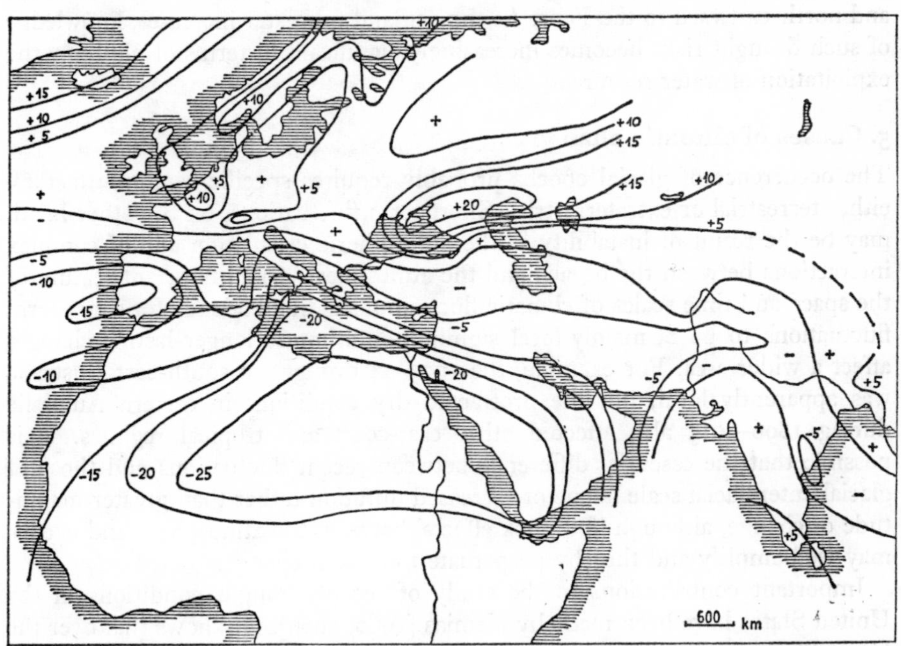

Fig. 11.1.4 Annual precipitation anomalies 1881–1910 to 1911–40, expressed as percentage deviations from the mean 1881–1910 (From Butzer, 1957).

Within the general trend of annual or seasonal averages less obvious, but none the less important, changes may occur. For example, the mean annual totals at Madras for 1813–80 and 1881–1940 are identical, but while the minor rains of May–July decreased after about 1890, the main October–December rains showed the opposite trend. Howe *et al.* [1966] demonstrate a marked increase in frequency of storm rainfall (daily totals of at least 2·5 in.) in mid-Wales during 1940–64 compared with 1911–40, while annual totals show only small changes. This caused an increase in flooding hazard. In New Mexico, Leopold [1951] found a steady increase in the frequency of daily rainfalls of less than 0·5 in. from 1850 to 1930 or 1940, and a subsequent decrease, without any variation in total rainfall. Such changes in semi-arid areas can be critical for plant growth, runoff, and erosion processes.

Precipitation fluctuations may assume economic significance, even in Britain. The period December 1963–February 1964 was the driest over England and Wales for more than 250 years, and if the September–February means are considered the years 1962–3, 1963–4, and 1964–5 had the lowest triple total, 45·1 in. (114·5 cm) for England and Wales, since 1757–60, when 44·6 in. (113·3 cm) was recorded. Dry winters lead to acute soil-moisture deficits in the following spring and summer, and to the likelihood of water restrictions in the south-eastern half of the country. Rodda [1965] estimates a 35–40% chance of a theoretical soil-moisture deficit (assuming that the potential evaporation rate is maintained) of 5 in. or more during the summer, in the area from Hampshire eastward to Kent and north-eastward to the Fens. As the demand for water increases, knowledge of such drought risks becomes increasingly significant in terms of planning the exploitation of water resources.

5. Causes of climatic change

The occurrence of glacial epochs probably requires specific causative factors, either terrestrial or extraterrestrial. Short-term fluctuations, on the other hand, may be the result of instability in the atmospheric circulation and of complex interactions between the oceans and the atmosphere. One important feature of the space and time scales of climatic fluctuations is the tendency for short-term fluctuations to be of mainly local significance, whereas longer-lasting changes affect a wider area. For example, the 1939–44 drought in south-east Australia was apparently local, but the protracted dry conditions in eastern Australia during 1896–1915 also affected other east-coast and tropical regimes. It is possible that the essential difference between recent fluctuations and those of glacial–interglacial scale is one of increased duration rather than greater magnitude of change, although 'feedback effects' between the atmosphere and oceans may well amplify and thereby perpetuate the initial change.

Important contributions to the study of recent drought conditions in the United States have been made by Namias [1960, 1966]. He shows that over the Great Plains a warm, dry spring tends to be followed by a warm, dry summer, whereas a cold spring is likely to be succeeded by a cold, wet summer. The physical mechanism involved is not yet clear. Drought over the north-eastern United States during 1962–5 seems to have been caused by below-normal sea surface temperatures offshore in spring and summer. This leads to increased cyclonic activity over the zone of maximum sea-surface temperature gradient, thereby strengthening dry, cool, north and north-westerly airflow over land. Corroboration of this idea is shown by the fact that three wet years, 1951–3, were associated with positive sea-temperature anomalies.

Precipitation changes with a time-scale of 50–100 years appear to involve the atmospheric circulation over at least the major part of one hemisphere. Kraus and Butzer suggest complementary explanations for the precipitation changes in the tropics and subtropics outlined earlier. In the tropics there seems to have been a shorter wet season associated with a narrowing of the intertropical convergence zone. The subtropical high-pressure cells apparently intensified and expanded both polewards and equatorwards, while cyclones in the westerlies of the northern hemisphere tended to shift northward. This represents a more zonal circulation, a tendency which has been drastically reversed since about 1940. In Mexico, however, Wallen [1955] found evidence of a significant increase in

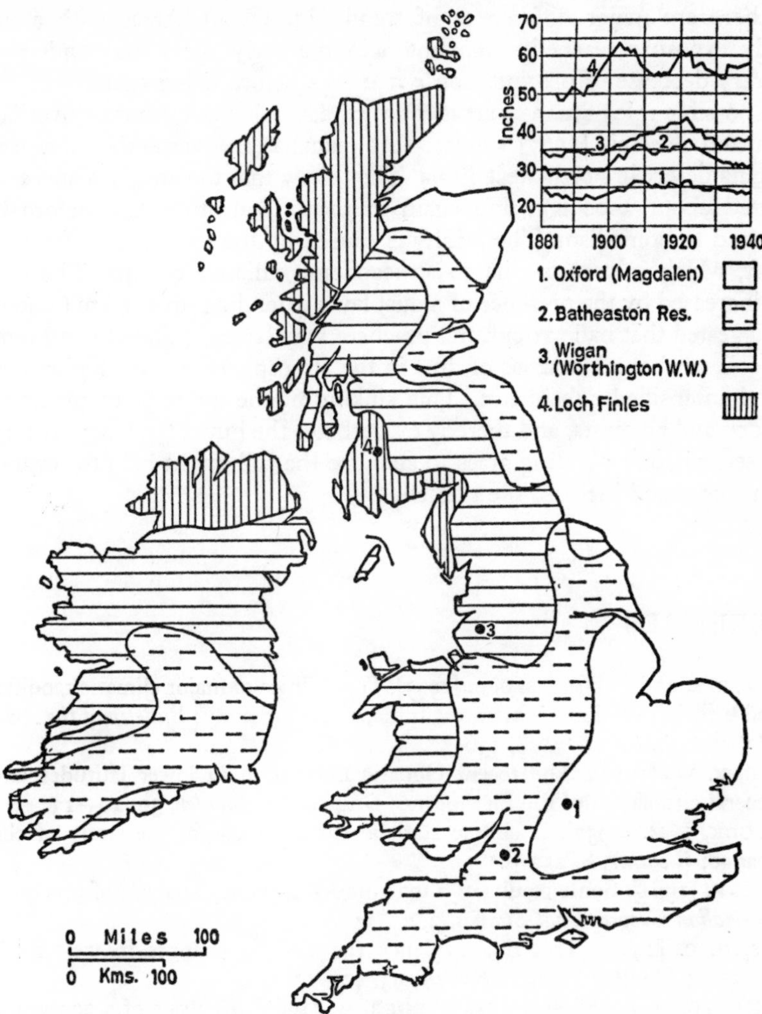

Fig. 11.1.5 Regions of annual rainfall fluctuations in the British Isles and their characteristic trends 1881–1940 (From Gregory, 1964).

annual, and especially July, precipitation from the 1900s to about 1930. This regional anomaly in the tropics need not invalidate the hypothesis of Kraus, but it does illustrate the necessity for cautious generalization. Similarly, Lamb [1967] demonstrates that regions exposed to prevailing winds from an adjacent ocean display a pattern of increased precipitation between about 1900 and 1940, corresponding to the period of strong wind circulation noted above, and a subsequent decline. Over England and Wales, for example, frequent westerly airflow, and therefore greater atmospheric transport of moisture, was experienced during 1910–40, with decadal means of precipitation for the 1910s and 1920s about 25% greater than in the 1880s. However, even in the British Isles there are major differences of trend (fig. 11.I.5). Areas with a north-westerly exposure reached a maximum in the early 1900s and underwent a temporary decrease of precipitation *c.* 1913–22 before rising again.

The possible role of man as an agent of climatic change cannot be overlooked. Bryson and Baerreis [1967] suggest that man has inadvertently extended the Rajputana desert in north-west India. They show that the area was successfully cultivated about 2000 B.C. and again in the fifth century A.D. before being abandoned to nomadism. The aridity is caused by the dominance of subsiding air aloft, which in turn is due to atmospheric radiative cooling. The cooling rate is increased by the presence of a dust layer extending up to 9 km (30,000 ft). It is suggested that bad agricultural practices may have facilitated wind removal of the soil, so that an increase of dust in the atmosphere led to a higher cooling rate and intensified subsidence. This sinking of the air reduces precipitation frequency and amounts, and thereby exacerbates the initial tendency. It remains to be seen whether planting grass to stabilize the soil might help to reverse the whole process and increase the rainfall.

REFERENCES

BRYSON, R. A. and BAERREIS, D. A. [1967], Possibilities of major climatic modification and their implications: Northwest India, a case for study; *Bulletin of the American Meteorological Society*, **48**, 136–42.

BUTZER, K. W. [1957], The recent climatic fluctuation in lower latitudes and the general circulation of the Pleistocene; *Geografiska Annaler*, **39**, 105–13.

CRADDOCK, J. M. [1957], A simple statistical test for use in the study of climatic change; *Weather*, **8**, 252–8.

DURY, G. H. [1964], Some results of a magnitude-frequency analysis of precipitation; *Australian Geographical Studies*, **2**, 21–34.

FRITTS, H. C. [1965], Tree-ring evidence for climatic changes in western North America; *Monthly Weather Review*, **93**, 421–43.

GIOVINETTO, M. B. and SCHWERDTFEGER, W. [1966], Analysis of a 200-year snow accumulation series from the South Pole; *Archiv für Meteorologie, Geophysik, und Bioklimatologie, A*, **15**, 227–50.

GREGORY, S. [1956], Regional variations in the trend of annual rainfall over the British Isles; *Geographical Journal*, **122**, 346–53.

HOWE, G. M., SLAYMAKER, H. O., and HARDING, D. M. [1966], Flood hazard in mid-Wales; *Nature*, **212**, 584–5.

JULIAN, P. R., and FRITTS, H. C. [1967], On the possibility of quantitatively extending precipitation records by means of dendroclimatological analysis; *International Association of Scientific Hydrology, General Assembly of Bern*, 243–50.

KRAUS, E. B. [1955], Secular changes of tropical rainfall regimes; *Quarterly Journal of the Royal Meteorological Society*, **81**, 198–210.

KRAUS, E. B. [1958], Recent climatic changes; *Nature*, **181**, 666–8.

LAMB, H. H. [1965], The early Medieval warm period and its sequel; *Palaeogeography, Palaeoclimatology, Palaeoecology*, **1**, 13–37.

LAMB, H. H. [1967], Britain's changing climate; *Geographical Journal*, **133**, 445–66.

LEOPOLD, L. B. [1951], Rainfall frequency: an aspect of climatic variation; *Transactions of the American Geophysical Union*, **32**, 347–57.

LEWIS, P. [1960], The use of moving averages in the analysis of time-series; *Weather*, **15**, 121–6.

MITCHELL, J. M., JR. *et al.* [1966], *Climatic Change;* World Meteorological Organization, Technical Note No. 79 (Geneva), 72 p.

NAMIAS, J. [1960], Factors in the initiation, perpetuation and termination of drought; *International Association of Scientific Hydrology, Publication No. 51*, 81–91.

NAMIAS, J. [1966], Nature and possible causes of the northeastern United States drought during 1962–65; *Monthly Weather Review*, **94**, 543–54.

RODDA, J. C. [1965], A drought study in south-east England; *Water and Water Engineering*, **69**, 316–21.

SMITH, M. P. [1965], Crisis looming on water shortage; *The Times*, 15 July, 11

VERYARD, R. G. [1963], A review of studies on climatic fluctuations during the period of meterological record; In *Changes of Climate*, Arid Zone Research, 20, UNESCO, (Paris), 3–15.

WALLÉN, C. C. [1955], Some characteristics of precipitation in Mexico, *Geografiska Annaler*, **37**, 51–85.

Index